ANCIENT EARTH, ANCIENT SKIES

Ancient Earth, Ancient Skies

THE AGE OF EARTH AND ITS COSMIC SURROUNDINGS

G. Brent Dalrymple

STANFORD UNIVERSITY PRESS
STANFORD, CALIFORNIA
2004

Stanford University Press
Stanford, California

© 2004 by the Board of Trustees
of the Leland Stanford Junior University.
All rights reserved.

Printed in the United States of America
on acid-free, archival-quality paper

Library of Congress Cataloging-in-Publication Data

Dalrymple, G. Brent.
Ancient Earth, ancient skies : the age of Earth and its cosmic surroundings / G. Brent Dalrymple.
p. cm.
Includes bibliographical references and index.
ISBN 0-8047-4932-9 (cloth : alk. paper)—
ISBN 0-8047-4933-7 (pbk. : alk. paper)
1. Earth. 2. Solar system. 3. Meteorites. 4. Cosmology. I. Title.
QB631.D34 2004
525—dc22 2003024023

Original Printing 2004
Last figure below indicates year of this printing:
13 12 11 10 09 08 07 06 05 04

Typeset by Heather Boone in 11/14 Garamond.

Contents

TO THE MEMORY OF ALLAN COX

Preface

I decided to write this book in 1983. Better late than never, I suppose. The impetus was the Arkansas "creationism" trial, *McLean v Arkansas*, held in December 1981 in Little Rock (529 F. Supp. 1255, E.D. Arkansas, 1982). Earlier that year the legislature had passed, and the governor had signed, Act 590, the Balanced Treatment for Creation-Science and Evolution-Science Act. This ill-advised law required that "creation science" be taught in the public schools of Arkansas whenever "evolution science" was taught. Act 590 defined creation science, an oxymoron if there ever was one, to include the concepts of a young Earth, the instantaneous creation of all living "kinds," the denial of evolution, and numerous corruptions of science required by these tenets. "Evolution science," as defined by the Act, included much of biology, geology, astronomy, chemistry, and physics as they are currently understood by scientists. The Arkansas law was challenged as a violation of the First Amendment to the U.S. Constitution and overturned in federal district court after a trial that attracted nationwide attention. (For an entertaining account, see Gilkey, L., *Creationism on Trial*, Minneapolis: Winston Press, 1985.)

I was called by the ACLU to testify in that trial as an expert witness on radiometric dating of rocks, the breadth of geologic time, the age of planet Earth, and the fallacies of "creation science" in these areas. Arkansas and the creationists lost that lawsuit, and the State of Louisiana lost the challenge to a similar law, in which I also participated, shortly thereafter (*Aguillard v Treen*, 634 F. Supp. 436, E.D. Louisiana, 1985). The U.S. Supreme Court affirmed the Louisiana decision in 1987, making such equal time laws invalid forever (*Edwards v Aguillard* 482 U.S. 578 1987).

As a result of the Arkansas and Louisiana cases (and another in California), I received numerous invitations to speak and write on the age of Earth and the fallacies of the creationist views on that subject. I was often asked if there was a current book or review article on the subject. Unfortunately, there was none, so I decided to write one. What I had in mind was a book that explained, in a simple and straightforward way, the data and logic that have led scientists to conclude that Earth and the other parts of the Solar System are 4.5 billion years old, and the Universe older still. (There. I've answered the question posed by the book's title, so those interested only in the result need read no further. If, however, you are interested in *how* the answer was arrived at, please read on.)

What I ended up with five years later was a book I liked a lot, but it was more scholarly and detailed than I had originally intended. Entitled *The Age of the Earth*, it is 474 pages long, complete with the relevant mathematical equations, references, and notes. Stanford University Press published it in 1991. The book filled an obvious need and is still in print, but it is sufficiently detailed that it is of interest primarily to people with some knowledge of science.

The book you are now reading is updated, abridged, and greatly simplified compared to the original volume. It is free of the equations, many of the references and notes, and much of the detail contained in the longer version. Scientific terms are explained when they first occur in the narrative, where practical, and a glossary is included in the back of the book. I hope it will prove to be interesting and informative to nonscientists who want to know about the evidence for the age of Earth, the Solar System, the Milky Way Galaxy, and the Universe. Readers of this book who wish more detail or documentation than presented here are referred to the 1991 version, which covers all of the same ground and more, with the exception that it is current only to about 1988.

The purpose of this book, then, is to explain how scientists have deduced the age of planet Earth, the Solar System, the Milky Way Galaxy, and the Universe. The Universe is a large, old, and complicated place. Earth and the other bodies in the Solar System—the planets, the asteroids, the comets, and so forth—have endured a long and sometimes violent history, the events of which have largely obscured the record that scientists wish to unravel. Although in detail the journey into Earth's past requires considerable scientific skill, not to mention access to modern and complex technology, the basic story is not so complicated that it cannot be understood by anyone who has the desire to learn and understand a few things about the evidence.

Many thanks to Jo Ellen Parker and Patrick Muffler, who reviewed the entire manuscript, and to Sam Bowring and Marc Davis, who reviewed Chapters 4 and 10, respectively. Their many thoughtful suggestions greatly improved the manuscript. The excellent staff of the Valley Library at Oregon State University was most helpful in making accessible, often quickly and in electronic form, copies of the many important and relevant scientific papers published since 1988, and I thank them for their help. Several colleagues generously provided prepublication copies of their papers, for which I am especially grateful. Finally, my thanks once again to my wife and best friend Sharon, who not only puts up with my time-consuming projects but also has provided love, companionship, support, and understanding for more than four decades.

G. Brent Dalrymple
Corvallis, Oregon

About the Author

G. Brent Dalrymple received his A.B. from Occidental College in 1959, with a major in geology, and his Ph.D. in geology from the University of California at Berkeley in 1963, where he learned how to date rocks using isotopes. He served 31 years as a Research Geologist with the U.S. Geological Survey in Menlo Park, California. In 1994 he moved to Oregon State University, to become Dean of the College of Oceanic and Atmospheric Sciences. He retired in 2001 and is currently Emeritus Dean and Professor at OSU, but spends most of his time these days engaged in woodworking, photography, downhill skiing, and golf, as the seasons dictate.

Brent's primary research interests include the development and improvement of isotope dating techniques and their application to a variety of geologic and geophysical problems. His most noteworthy accomplishments include the determination of the time scale of reversals of Earth's magnetic field for the past 4 million years, a research collaboration with colleagues Allan Cox and Richard Doell that led to the Theory of Plate Tectonics, and the demonstration that the Hawaiian-Emperor volcanic chain was formed by motion of the Pacific Plate over a fixed source of lava in Earth's mantle. His most recent research includes the timing of the large impacts that created the prominent lunar basins, some of which form the face of the "Man in the Moon."

In 1975 he was elected a Fellow of the American Geophysical Union, a professional society for which he served as President from 1990 to 1992. He was elected to the American Academy of Arts and Sciences in 1992 and to the National Academy of Sciences in 1993. He received an honorary Doctor of Science degree from Occidental College in 1993 and the Public Service Award of the Geological Society of America in 2001.

His bibliography includes more than 150 papers in scientific journals, a textbook on potassium-argon dating (coauthored with M.A. Lanphere), and a book entitled, *The Age of the Earth*, published by Stanford University Press in 1991.

Brent and his wife Sharon, a retired mathematics teacher, live in Corvallis, Oregon. They have three daughters: Stacie, the Senior Director of Pharmacology for a biotechnology company; Robynne, the Senior Director of Americas Revenue for a software company; and Melinda, a Lieutenant Commander on the faculty of the U.S. Coast Guard Academy. Brent and Sharon consider these three beautiful and talented people their greatest accomplishments.

The chessboard is the world, the pieces are the phenomena of the universe, the rules of the game are what we call the laws of nature. The player on the other side is hidden from us. We know that his play is always fair, just, and patient. But also we know, to our cost, that he never overlooks a mistake, or makes the smallest allowance for ignorance.

—T. H. HUXLEY, 1868, *A Liberal Education*

When the heavens were a little blue arch, stuck with stars, me thought the universe was too straight and close: I was almost stifled for want of air: but now it is enlarged in height and breadth, and a thousand vortices taken in. I begin to breathe with more freedom, and I think the universe to be incomparably more magnificent than it was before.

—BERNARD LE BOVIER FONTENELLE, 1686

ANCIENT EARTH, ANCIENT SKIES

CHAPTER ONE

Sleuthing the Past

Four and a half billion years, the number that represents the current estimate of the age of Earth, is so large, so far outside of our normal, everyday experience that it is difficult to comprehend. If a piece of string an inch long represents one year, for example, then a 6-foot length is about equivalent to the average lifetime of a U.S. citizen. The length of a string representing all of recorded history would be slightly more than half a mile, but a string representing 4.5 billion years would be 71,023 miles long and would wrap around Earth nearly three times!

If a period of time measured in billions of years is difficult to grasp, the implications of such a temporal abyss are even more elusive. Think of all the things that have happened during your lifetime, and then try to imagine how many things might have happened everywhere in the Universe in 4.5 billion years. Stretches the mind, doesn't it?

As staggering as these numbers may seem, the evidence clearly shows that Earth's age is, in fact, 4.5 billion years and the Universe, including our Milky Way Galaxy, is about three times older. Yet humans are relatively recent inhabitants of the planet, the merest of specks on a timeline of Earth's history, and we have witnessed only an infinitesimally small percentage of what has happened since Earth formed. How, then, can scientists peer back through vast reaches of time, figure out so many things about the history of Earth and the Universe, and determine an age for Earth with an accuracy of better than 2%? It's a good question, and it deserves a good answer.

What Is Science and How Does It Work?

Although people have probably wondered about the mysteries of nature for millennia, science is a relatively recent field. Prior to the seventeenth century, most civilizations attempted to understand the physical world by appealing to religion. Things that appeared to be mysterious—such as eclipses of the Sun, lightning storms, diseases—were attributed to actions of the gods and often viewed with superstition and fear. The scientific and intellectual revolution that characterized the Age of Enlightenment, which developed in the seventeenth and eighteenth centuries in Western Europe, changed that dramatically. The leaders of this movement—including Kepler, Galileo, Bacon, Descartes, Newton, Boyle, and Lavoisier, to name a prominent few—began to seek an understanding of the physical world through observation, experiment, and the careful formulation of logical conclusions. In doing so, they freed the intellect from the bondage of medieval theology, made curiosity about nature legitimate, and invented what we know today as science.

Twentieth-century science is a vast, complex, and often daunting enterprise conducted by scientists working for universities, governments, and businesses. Scientists are the caretakers of an enormous system of knowledge that has accumulated and evolved over several centuries, and it is being constantly modified, corrected, and expanded. The purpose of some scientific work is to answer specific questions about matters of immediate importance. What are the long-term effects of certain chemicals on the human body? How can batteries be made that will make electric automobiles practical? How can cancer be prevented?

Much of the efforts of scientists, however, are directed simply at expanding our understanding of nature. Science is the activity of thousands of curious people who want to discover how nature works. Science does not exist, however, just to provide amusement and employment for scientists. On the contrary, even basic research that may seem to have no immediate application is of immense value to society. The extensive pool of accumulated scientific knowledge, to which new discoveries are constantly being added, is a valuable resource for others who seek to develop new products, solve technological problems, and improve our living conditions.

Like most organized forms of human activity, science has rules. Just as in baseball, the rules of science define the boundaries of the playing field and the types of moves and strategies allowed in the game. The rules of science are not written down in a formal code, in the manner of civil law or the rules of baseball, but there is an extensive body of literature about them, and they are well understood and carefully followed. They were developed and are followed out of necessity, because without them science would not advance.

The rules are quite simple, and you use some of them every day without realizing it. For example, when you go to bed tonight you fully expect that the Sun will rise tomorrow morning. In arriving at this conclusion you have used inductive reasoning, that is, the drawing of general conclusions from specific observations, one of the primary tools of science. Let's briefly explore some of the more important rules of science, especially as they apply to the processes of probing into the mysteries of the past.

THE SUBJECT IS NATURE

This is probably the most important rule of all, for it determines what questions scientists can and cannot ask (and answer) and provides the fundamental basis for understanding scientific conclusions. Failure to understand this rule is one of the primary reasons that science is so often perceived to be at odds with other fields of intellectual endeavor, especially religion. Put simply, science involves the study, description, and classification of nature and natural processes. It deals only with the world of physical phenomena. Politics, art, baseball, and affairs of the human spirit, such as religion and philosophy, are not included because these are subjects with which science simply is not equipped to deal.

This is not to say that science does not overlap with other disciplines. On the contrary, science commonly peeks and pokes into any field where its unique methods of investigation might bring clarification. Science can explain the physics of a curve ball, but not why baseball is so popular. Science can explain why the mixture of certain colors produces other colors, but not why we find a painting by Rembrandt or Van Gogh so beautiful. Science can determine the age of the Shroud of Turin (A.D. 1325 ± 65 years) but has nothing to offer on its religious significance.

Science has a unique role in finding natural explanations for things regarded by some as mysterious—flying saucers revealed as electrical discharges and ordinary rocket launches, supposed paranormal phenomena exposed as fraud, "creation science" revealed as thinly disguised religious apologetics. This restriction to nature and the neglect of other matters does not mean that science is either inferior or superior to other fields, only that it is different. Its limitation should not be considered either a weakness or a strength, but merely an essential attribute. All it means is that science has its limits, just as there are limits to the size of a football field and the diameter of a golf ball.

Because science deals solely with observable phenomena that occur in the physical world, there are some questions that science can answer and others it cannot. Science can be expected to provide a reasonable estimate for the age of Earth and to describe the sequence of physical events that sculpted the world we now inhabit. But it cannot answer the questions of why we exist, who created the Universe, or even if there was a "who" involved in the process. If those questions have answers, they must come from the fields of religion and philosophy, not science.

CAUSE AND EFFECT

The premise that for every effect there must be a cause may seem so obvious that it appears unnecessary to state it, but it is the very basis for asking questions about nature. The law of cause and effect has been observed to be true in so much of our experience that it is only reasonable to assume it holds for all of nature and for all time. It tells us that for every observable condition, it is profitable and worthwhile to seek a rational and natural cause, and that without understanding the cause, one can never really understand the effect. Perhaps more fundamentally, it says that answers to questions about nature do exist, and that there is every chance that we can find them if we are clever and persistent enough. Scientists presume that there is a reason for everything that can be observed in nature, and, conversely, that nothing happens in nature without there being a rational cause that can eventually be determined by the careful application of experiment, observation, and inductive reasoning.

EXPERIMENTS DEMYSTIFIED

A frequent barrier to understanding how scientists decipher the past is a common misconception of the definition and role of the experiment in science. Most people realize that experiments play an important role in scientific discovery. Many think of an experiment, however, exclusively as a test that can be done in the laboratory and repeated at will. If this is so, they ask, then how can scientists experiment with the past to discover what happened? Sometimes the question takes a slightly different form. Since there were no people millions of years ago to observe and record events as they occurred, how can scientists confidently say what happened so long ago?

Webster's dictionary defines an experiment as "a test or trial of something; specifically, any action or process undertaken to discover something not yet known or to demonstrate something known." The key word here is *process.* Included within this definition are three distinct types of experiments, each involving a different process. The real-time experiment is the familiar one. Common in physics and chemistry, the real-time experiment involves repeatability at will and the careful control of, and the ability to vary, experimental conditions. An example is an experiment to determine the temperature at which a particular type of rock melts. The scientist melts samples of the rock in the laboratory and carefully measures the melting temperature. By varying some of the conditions of the experiment, the scientist can determine how the melting temperature is affected by pressure, the presence of different amounts of water, and so forth. Any step of the experiment can be repeated at any time by anyone who desires to question or verify the results.

In the derived or historical sciences, including geology and astronomy, a second type of experiment, the observational experiment, is common. This type of experiment depends on observing the effects produced by both past and present natural events. The experimental conditions cannot be varied at will by the scientist, as they can in the real-time experiment, but an observational experiment can be repeated by different observers, at different localities, and at different times in order to verify results. The type of observation can also be varied. Consider an experiment to determine the climate at some particular time in the past. If a large proportion of the rocks formed during that time are identical to the deposits of modern glaciers, and the fossil plants and animals preserved in the rocks of that time resemble modern forms that today live only in cold climates, then the scientist would conclude that the climate in the region when the rocks formed was probably cold. Others could check the results at any time by repeating the scientist's original observations and by adding new observations from other localities.

The third type of experiment, the thought experiment, takes place solely within the mind and is an exercise in pure logic. The conditions can be varied at will and the experiment is repeatable, but there are no physical measurements or observations. This type of experiment involves formulating an imaginary but plausible set of initial conditions, and following the resulting sequence of events and their effects through to their logical conclusion within the confines of the relevant natural laws. Albert Einstein often conducted thought experiments in his research on relativity. One well-known example is his experiment involving twins, one of whom embarks on a round-trip journey to a nearby star at a speed approaching that of light. Einstein's theory of special relativity requires that time slows down for anything traveling at a velocity near the speed of light. The result is that the traveling twin observes that his trip takes less than a day, whereas the twin on Earth observes that his brother has been gone for decades. Both twins age according to their own frame of reference, and at the completion of the trip the twins are no longer the same age. Although this experiment was done entirely in Einstein's mind and involved no physical measurements, the results can be verified at any time by anyone with a sufficient knowledge of physics.

All three types of experiments are equally valid and powerful tools of science, and often two or even three will be brought to bear on a single topic of research. But it is the observational experiment, combined with the law of cause and effect and sometimes assisted by real-time experiments, that is the key to discovering the history of Earth, the Solar System, and the Universe. Thus, it is not necessary to have been there in order to reach valid conclusions about past events in nature, even though the events may have occurred millions or billions of years ago.

NO MAGIC, PLEASE

While the law of cause and effect encourages scientists to seek causative agents to explain the existence of observable effects, there are limitations on the kinds of explanations that are acceptable. A fundamental premise of science is that natural laws do not change with time. We presume that the laws describing the properties and behavior of matter and energy today operate everywhere in the Universe and have operated throughout the history of the Universe. Where scientists have in the past thought that some anomaly was defying the laws, they later learned that they had not been examining the anomaly properly or that the original law was inadequate—and then understanding of the scientific law changed accordingly.

If natural laws are constant and predictable, then it follows that supernatural agents may not be invoked in science; magic, witchcraft, or intervention by a supreme being are excluded as possible causes. This does not mean that a supreme being does not exist or that Earth was not created by some miraculous event, only that such an explanation is forbidden in the world of science. Why is this? If science were not restricted to natural explanations, there would be little reason to seek them, for everything could instead be explained easily by calling upon supernatural acts. Admittedly this would save a lot of effort and expense, but it would also result in a rather unpredictable and useless science. Science takes as one of its starting points the premise that nature is decipherable, and that it is reasonable and profitable to ask questions about the history of the Universe, including the age of Earth.

SIMPLIFY, SIMPLIFY

Science always seeks and adopts the simplest of all possible answers that are consistent with the facts, unless there is good reason to do otherwise. It is easy to make up complicated explanations, but scientists learned long ago that the most efficient way to proceed is to find the simplest explanation and add complexity only when required to satisfy specific observations.

One of the primary objectives of science is to find new unifying laws and ordering principles (theories) to explain and describe complex phenomena. Examples are the laws of thermodynamics, which explain the relationships between heat and various other forms of energy; the law of conservation of momentum, which explains the resistance of bodies to changes in motion; the theory of evolution, which accounts for the observation that organisms change over time; the theory of relativity, which explains important relationships between mass, energy, space, time, gravity, and motion; and the theory of plate tectonics, which accounts for the motions of Earth's outer layers. Scientific laws and theories are discovered by the application of inductive reasoning to generally small sets of observations.

Consider the sunrise. If you expect the Sun to rise tomorrow, you have probably reached that conclusion by the application of observation, induction, and simplification, even though you may not have realized you were applying scientific reasoning to the results of an observational experiment. In our short lifetimes, we observe only a very small percentage of the trillion or so sunrises that have occurred on Earth. How, then, can we be so confident that tomorrow's sunrise will arrive on schedule? The key is logical consistency. Sunrises occur with such regularity and predictability that it is reasonable to presume that they will continue for the foreseeable future. Even the fact that the time of sunrise varies from day to day and from one location on Earth to another, or the knowledge that several billion years from now the Sun will consume its available energy and cease to function as it does now, does not shake our confidence because even those changes are predictable. Our experience has convinced us that the simplest conclusion—that the Sun will rise again tomorrow whether or not we are here to see it—is the most profitable because it is the one most likely to be correct.

We can also use inductive reasoning and simplicity in arriving at conclusions about the history of Earth. On the basis of observations we make today, and presuming the constancy of natural laws, we arrive at reasonable conclusions about events in the past, making them no more complicated than is necessary to account for the facts at hand. This process is responsible for the success of the historical or observational sciences, like geology, where real-time experimentation often is intrinsically impossible. For example, on the basis of our observations of sunrise—including the written historical record, which confirms that the phenomenon occurred throughout human history, and the fossil record, which shows that photosynthetic plants existed hundreds of millions of years ago—are we not entitled to conclude that sunrise has occurred regularly over all of geologic time? The answer is yes, unless there is some evidence to the contrary. Any other conclusion would be more complicated than necessary and would be inconsistent with the principle of the constancy of natural laws. It would also make scheduling of tomorrow's activities extremely difficult.

Consider, as another example, sand dunes. Numerous observations have shown that the sand grains found in modern dunes have surface characteristics unique to wind-blown sand. In addition, the sand in dunes is deposited in a sequence of inclined and curved layers, called cross-bedding, that is distinctly different from the cross-bedding found in sand deposited by water. If ancient rocks in which the sand shows all of the characteristics of sand in modern dunes are found, what is the logical conclusion about the origin of these rocks? By now you probably have the idea.

ANYONE CAN PLAY

Science is a game that is open to anyone who wishes to play and will agree to abide by the rules. One of the premises of science is that observations and experiments must be reproducible by others, and every result is subject to verification. This means that science is nonauthoritarian. No individual, committee, organization, or government dictates which conclusions are valid and which are not, and no conclusion or observation is accepted as scientific unless it is made and expressed in such a way that it can be confirmed by anyone who is able to repeat the process. This also means that conclusions are accepted as valid only by the consensus of knowledgeable scientists. There are no votes and no official pronouncements. Each scientist makes those decisions on his or her own. When a large majority of scientists are convinced that some finding is valid, then that finding becomes accepted by the scientific community more or less automatically, and it will be the working model until it is disproved or shown to be inadequate.

There are, to be sure, individual scientists and scientific organizations that are listened to more carefully than others because of their expertise and reputations for excellence. The U.S. National Academy of Sciences, for example, has a reputation for conducting studies on issues using panels of experts and a rigorous review process. As a result, their reports are taken very seriously by the scientific community and by the U.S. government. Some individual scientists are recognized as authorities in their fields because of the quality and depth of their research, knowledge, and experience. The reasoning and conclusions of these scientists may carry more weight than those of others. Anyone is free to disagree, publicly or privately, with any of these authorities, but if one does so, he or she should have some better data or more insightful reasoning to back up the objections.

Several things make scientific research fun. One is the thrill of discovery. Another is the excitement of competition. Competition is also why science is self-correcting and is a field in which there is a significant disincentive to cheat. No scientist can exclude others from a field of inquiry, so there are invariably many scientists working on similar things. If it appears that one scientist has discovered something important, it is a sure bet that others will take up the same or similar research path, check the results, and either confirm, extend, or disprove the discovery. To confirm or improve the finding is fun because one then becomes part of the discovery. It is also great fun to recognize what others have overlooked, and it is much less embarrassing for a scientist to correct his or her own mistakes publicly than to have someone else do it. What this all means is that important findings are checked and rechecked, sometimes over a period of many years, and that mistakes are eventually discovered and publicly revealed. Science is self-correcting by design and is continuously changing as new knowledge is acquired.

NOTHING IS FOREVER

A final attribute of science that often causes confusion and misunderstanding by nonscientists is that everything science discovers is considered by scientists to be tentative. This is why scientists often use so many qualifiers when asked to explain some recent finding or conclusion to the public. This situation is not helped by some of the words scientists use—words like hypothesis, law, and theory—to describe the structure of science. The confusion comes because the general public uses these words somewhat differently than scientists do. *Hypothesis* is probably the least misunderstood. A scientific hypothesis is a tentative model advanced to explain some usually incomplete set of observations. Nonscientists use the term similarly. Often a scientist will advance several hypotheses to explain the same set of observations. Next, these hypotheses are tested to see which ones hold up and which need to be discarded as unworkable.

The terms *law* and *theory*, however, are sources of considerable confusion. A scientific law is a concise statement of a relationship between phenomena that is invariable under a given set of conditions. Laws often have a simple mathematical expression, such as $E = mc^2$, which describes exactly the relationship between energy (E), mass (m), and the speed of light (c), but there are also laws that do not. The law of superposition, so important in geology and first articulated by the Danish physician and naturalist Nicolaus Steno in 1669, simply states that in any undisturbed sequence of sedimentary strata, any individual stratum is older than the one above it and younger than the one below.

But scientific laws are not always as inviolable as the word implies. Sometimes they are proved wrong and discarded. Sometimes they are proved wrong but still used under certain conditions. Newton's law of gravity is a good example of the latter. It has a simple mathematical expression that treats gravity as a kind of force that attracts bodies toward each other. Although this works in a wide variety of situations and is still commonly used, it is not exactly true. Newton's law fails when very large masses, like stars and galaxies, are involved. Albert Einstein's theory of general relativity reveals that gravity is not a force but instead that mass causes curvature in space. As a result, objects are not actually attracted to each other, but instead they fall toward each other because of the curvature, much like two eggs roll to the bottom of a bowl. Relativity accounts for the observations that a clock runs more slowly (time slows down) in the presence of large masses compared with negligible masses, and it also explains why light, which has no mass, is deflected by large gravitational fields. There are other consequences and observations that Newton's law does not account for but that the theory of general relativity does. When appropriate, however, scientists still use Newton's law because it is a lot easier calculation to make and accurate enough for everyday situations on Earth.

Perhaps the most misunderstood term is *scientific theory.* To many nonscientists, a theory is nothing more than an idea, a guess, a belief, or a hypothesis. To a scientist, however, a theory is one of the most powerful statements that science can make about how the natural world works. A scientific theory is a logical and unifying structure of ideas that accounts for a large body of observations and, therefore, explains something important about nature. Examples are the theories of special and general relativity, the theory of evolution, and the theory of plate tectonics. Much of physics does not make any sense unless viewed within the framework of relativity. The theory of evolution and the theory of plate tectonics have the same importance to biology and geology, respectively. A theory is the closest that science can come to the "truth." This is why scientists were so upset when President Ronald Reagan said that evolution was "only a theory." He was equating an important scientific theory, for which there is convincing evidence and that is universally accepted by knowledgeable scientists, with a hypothesis or a guess, and he was dead wrong.

Despite the confidence scientists have in current theories and laws, none of them is absolutely guaranteed to endure. Science arrives at its description of nature by a series of increasingly better approximations. New findings may modify scientific facts, laws, and theories and lead to new and quite different conclusions from those previously thought to be true. Occasionally, when overwhelming evidence is brought to bear, a whole field of science may undergo a revolution of sorts—a "paradigm shift." The discovery of atomic structure around the turn of the twentieth century did that for chemistry and physics, just as the discovery of plate tectonics in the 1960s did for geology and Charles Darwin's 1859 descent with modification, now known as the theory of evolution, did for biology.

Revisiting the Inquiry

I have now answered the questions posed at the beginning of the chapter and shown, I hope, that it is not unreasonable for scientists to presume to answer the question about the age of Earth. It is merely an exercise in the application of inductive reasoning to careful observations of the present-day world and applied to the past on the presumption that there is a consistency and predictability to natural laws. It is also the result of the dogged and necessary premise that miracles or supernatural agents do not play any direct role. As I will show, the currently accepted age for Earth is a conclusion based on a breadth and depth of evidence that, at present, can be interpreted in no other rational way. The data, the reasoning, and the conclusions are so persuasive in this case that no serious scientist believes that the age so calculated could be dramatically in error. This does not mean that the age of 4.5 billion years is absolutely and forever firm. Future findings may cause scientists to revise the number slightly, or perhaps even drastically, but for now it is the best and only reasonable conclusion that science can offer. By the time you finish reading this book, I think you will agree it is a fair one.

Why Bother?

It is likely that humans have wondered about the age and origin of their surroundings since they first developed the capacity for abstract curiosity. How and when was the world created? Theologians, philosophers, and scientists have been searching for satisfying answers to these questions for thousands of years. Only within the last half century, however, has the age of Earth been known with any reasonable degree of certainty. Even then, the answer emerged only after centuries of thought, observation, and experimentation.

Why bother? Two reasons come immediately to mind. The first is philosophical. History makes it abundantly clear that people have struggled with the question of their place in the Universe since the beginning of organized thought. Was the Universe created by a god specifically for us, or are we a minor result of natural processes shaping a Universe of unimaginable dimensions over seemingly infinite time? Clearly, there is a certain comfort and security in the former explanation, and throughout most of recorded history, Western thought has focused on the idea that *homo sapiens* is central to a grand and purposeful plan conceived by a supreme being. Until Copernicus, Galileo, Kepler, and their successors showed otherwise, it was universally held that Earth, and by implication humankind, was at the very center of the Universe. Even after man discovered that his place in the Universe was not geometrically central, he still clung to the belief that his timing was. Indeed, some creationists still prefer to believe that the Universe was created specifically for humans and that creation predated the debut of humans by only a few days.

We now know, of course, that Earth revolves about a rather ordinary star located among billions of similar stars in a rather ordinary galaxy that occupies an ordinary position among billions of other galaxies in the Universe. Humankind is neither located in the center of the Universe, nor the center of the Milky Way Galaxy, nor even in the center of the Solar System. We also know that we humans are newcomers on the scene, having appeared so recently that our existence occupies but a fleeting moment in the vast span of time since the Universe began. Thus, knowledge of the age of Earth and its surroundings puts our lives in perspective and gives us all a better idea of our physical place in the Universe.

The second reason is scientific. We have always sought information about our physical surroundings to satisfy our curiosity and to add to the pool of scientific knowledge from which we draw both intellectual and material satisfaction in the form of new ideas, new understandings, new directions for future research, new technology, and new inventions. Thus, the age of Earth is simply one more interesting thing to know in the array of information that scientists have gathered in the quest for an increasingly accurate description of the Universe. Curiosity may well have killed the cat, but there can be no doubt that the curiosity of scientists has immeasurably enriched us all in numerous ways.

CHAPTER TWO

The Birth of the Universe, the Galaxy, the Solar System, and Earth

Before turning to the age of Earth and related subjects we'll briefly explore the sequence of events that led to the formation of the Universe, the Milky Way Galaxy, the Solar System, and planet Earth. We will also consider what is meant by an "age" for Earth.

The concept of a single, unique age for Earth is not entirely valid because Earth did not appear instantaneously. Like all physical objects, it was "born" through a complex sequence of events over some interval of time, so the selection of a particular event or stage of evolution in Earth's early history to represent the time of its birth becomes important. Even then, the event we may wish to choose may not be accessible—evidence of its existence may no longer exist, may be out of reach, or may be undiscovered. Alternatively, it may be accessible but undatable by present technology. These difficulties might be very serious and render any age for Earth meaningless if the interval of time over which the planet formed was long compared to its age. As it turns out, however, Earth's birth process occurred very long ago (about 4.5 billion years) over a relatively short interval of time (less than 0.1 billion years), so the errors introduced by these difficulties are very small.

The Origin of the Universe

The physical processes that gave birth to the Milky Way Galaxy, the Solar System, and Earth were set in motion billions of years ago by an event known as the Big Bang. The first evidence for this singularly important event, which marks the beginning of the Universe we know, was discovered in the 1920s by the astronomer Edwin P. Hubble. Hubble and his colleagues, working at the Mt. Wilson Observatory in southern California, observed that all of the visible, distant galaxies are moving away from each other. In 1927, Georges Lemaître, a Belgian astronomer and cosmologist then working at the Massachusetts Institute of Technology, proposed that this interesting observation was most easily explained if the Universe started at a definable time in the past with a violent expansion of matter and energy that was originally highly compressed and intensely hot. This, then, was the conception of the idea now known as the Big Bang, a name coined by the British astronomer Sir Fred Hoyle in a 1949 radio broadcast. Subsequent observations have confirmed Hubble's original findings and have added important new evidence that confirms Lemaître's hypothesis. (Hubble's findings are discussed further in Chapter 10.)

Probably the most important confirmation came from a discovery in 1965 by Arno Penzias and Robert Wilson, two engineers working for Bell Laboratories. They found a background radiation, called the *cosmic microwave background*, that pervades the Universe and has all of the characteristics it was predicted to have if the Universe started with a Big Bang and has been expanding and cooling ever since. This was a grand observational experiment if there ever was one, and Penzias and Wilson were awarded the Nobel Prize in Physics for their discovery. The Big Bang is now the accepted scientific theory for the origin of the Universe—indeed, there are no credible, competing theories—and although much of what must have happened during such an astounding event is known, many of the particulars are still to be discovered.

What existed before the Big Bang is not known and may never be discovered, for there is no physical evidence from previous times. It is not even certain that there was a previous time. Whether the present Universe is the first universe or just the most recent in an infinite sequence of universes is a question that science, at present, has no way of answering—but it is, at least, a very interesting question.

The physical processes and events during and shortly after the Big Bang are subjects of much current research and some are poorly understood, but the early history of the Universe is now thought to be approximately as follows. At the very beginning, the temperature and density of the Universe were infinite, and it was filled with light of tremendous energy. The Big Bang initiated expansion (or, more properly, inflation), that is, a decrease in density accompanied by cooling, both of which are continuing to this day. Within the first 10 seconds or so after the Big Bang, the principal subatomic particles—positively charged protons and neutral neutrons (no electrical charge), which together form the nuclei of atoms, and negatively charged electrons, which constitute the outer shells of atoms—were created. In addition, the fledgling Universe cooled to a temperature of less than 1 billion degrees on the Kelvin scale (water freezes at 273 K, and at 0 K, called absolute zero, there is a complete absence of heat). By the time 100 seconds had passed, the temperature of the Universe had decreased enough to allow some neutrons and protons to combine to form the nuclei of helium atoms. It was not until about 300,000 years later, however, that the temperature had decreased to 10,000 K and the first simple atoms, consisting mostly of hydrogen with some helium, formed. The subsequent history of the Universe involved continued cooling, continued expansion and decreases in average density, and the assembly of matter into mostly gas with some dust, followed by condensation of the dust and gas into galaxies, stars, and planets.

Of the events in the birth of the Universe, only the age of the Big Bang itself can be estimated from physical measurements. Fortunately, the Big Bang is the one event that best represents the time of origin of the Universe. Unfortunately, the errors in determining this age can be large. In the past, values have ranged from 7 to 20 billion years, but current estimates are in the range of 13 to 15 billion years. The event that best represents the origin of the Universe can be dated, but so far not with high accuracy.

The Origin of the Galaxy

As the Universe expanded, irregularities in the distribution of the dust and gas developed, causing some of this material to collect into irregular clouds of matter. In a few cases, internal gravitation caused the cloud to contract, forming a proto-galaxy, a mass of rotating matter that would eventually become a coherent group of stars known as a galaxy. One such proto-galaxy was the ancestral Milky Way Galaxy, sometimes simply referred to as the Galaxy, of which our Solar System is a part. As contraction within the proto-galaxy proceeded, turbulence caused it to rotate and the gas cloud to fragment. Some of the individual fragments near the center of the Galaxy contracted to form the first stars. The more massive stars evolved quickly and exhausted their nuclear fuel; each ended its existence in a colossal explosion known as a supernova, which produced heavier elements. The matter from these supernovae was returned to the galactic pool, where it was available for incorporation into new stars. Most of this matter was gaseous, but some of the heavier elements, such as silicon, iron, and magnesium, condensed to form more dust. As the Galaxy continued to evolve, rotation caused it to flatten into a disk with a central bulge and eventually to form spiral arms (Figure 2.1), even as new stars were formed. All available evidence indicates that the Milky Way Galaxy continues to evolve. Stars are still forming in the core and spiral arms where the density of matter is highest, running their course, and dying. Those that perish as supernovae contribute their constituents to the formation of new stars.

In contrast to the beginning of the Universe, there is no single, clear-cut event that marks the beginning of the Milky Way Galaxy. Of the several events discussed above, which best represents the Galaxy's birthday—the formation of the proto-galaxy, the appearance of the first stars, the flattening of the Galactic disk? The choice of any of these would be arbitrary because none is a discrete event; each one is part of a continuous process. As a practical matter, the choice is limited because the ages of only three events can be determined. Some of the oldest stars in the Galaxy are found in globular star clusters, which occur within a halo that surrounds the central part of the galactic disk. The ages of these stars can be calculated by comparing their stage of development to theoretical rates of star evolution. It is also possible to estimate the ages of the oldest white dwarfs, which are former stars, in the Galaxy. Finally, the approximate age of the Galaxy can be estimated from observations of the abundances of the isotopes of certain heavy elements. These three methods and their results are discussed further in Chapter 10. For the moment, suffice it to say that none of these age estimates is as precise as we would like, but they are consistent with the calculated age of the Big Bang and indicate that the Milky Way Galaxy is about 11.5–14.0 billion years old.

The Origin of the Solar System

The disk of the Milky Way Galaxy is about 10^{17} kilometers in diameter, completes one revolution about its center every 200 million years, and contains something like 100 billion stars. (10^{17} is 1 followed by 17 zeros: 100,000,000,000,000,000.) About 26×10^{15} km from the center of the Galaxy, near the inner edge of the spiral arm named after the Orion Nebula, lies an average star called the Sun, which, as stars go, is unexceptional in size,

Figure 2.1 (Top) Spiral galaxies. The dusty spiral galaxy NGC 4414, as imaged by the Hubble Space Telescope, is about 60 million light years from Earth. (NASA photograph STScI-PRC99-25.) *(Bottom)* This Hubble Space Telescope image shows two spiral galaxies, NGC 2207 on the left and IC 2163 on the right, colliding. Many of the more massive galaxies, possibly including the Milky Way Galaxy, are thought to have formed by the collision and coalescence of smaller galaxies. (NASA photograph PR99-41.)

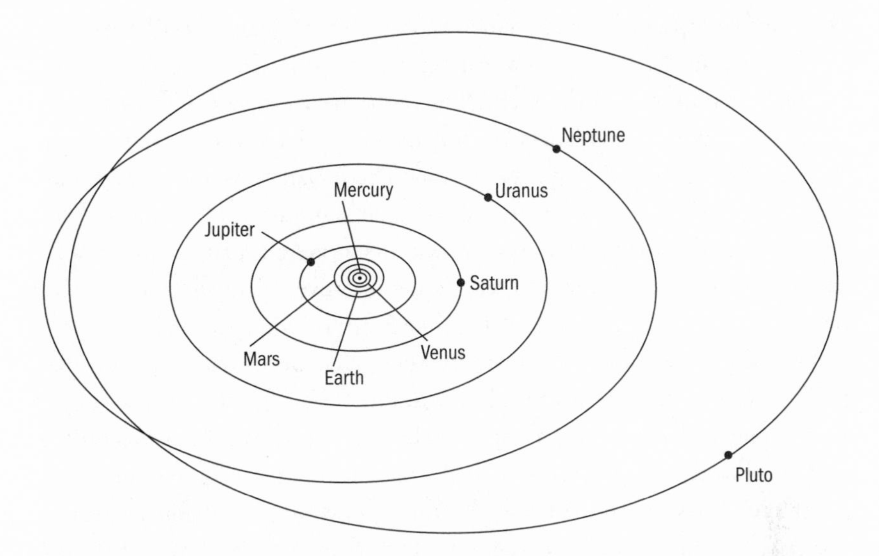

Figure 2.2 The Solar System drawn as if viewed at an angle. The orbits of the planets are shown at approximate scale and, except for Pluto, are nearly circular. Most of the asteroids (not shown) orbit in a belt between Mars and Jupiter.

brightness, temperature, color, and composition. Orbiting the Sun are nine planets, six of which have orbiting satellites, or moons, of their own. Various debris, including rocky bodies varying in size from small planets to large rocks (asteroids), rocks and pebbles of various sizes (meteoroids), comets, consisting primarily of ices, and dust, also orbit the Sun. The Sun and its various orbiting companions constitute the Solar System (Figure 2.2).

Are there planets elsewhere in the Universe? Yes. Planets are probably common features of many star systems. It seems likely that most stars have some small amount of debris left over from which planets might form. But planets are small and dark, and even those orbiting the nearest stars would be too far away to see visually, even with the most powerful of today's telescopes. Planets, however, do cause the stars they orbit to wobble as the planet and star orbit their common center of mass, and this wobble can be measured. Analysis of the wobble can reveal the mass of the planet, its distance from the star, and whether or not there is more than one planet present.

Since 1995, astronomers have found more than 70 stars with planets in orbit about them, and several of these stars have at least two or three planets.

For example, Epsilon Eridani, which is only 10 light years from Earth, has a planet of about the same mass as Jupiter, which orbits the star every 7 years. In 2001, astronomers at the Lick Observatory near San Jose, California, detected two Jupiter-sized planets orbiting the star 47 Ursa Major, a faint star in the Big Dipper about 51 light years away. These planets were the first found to move in nearly circular orbits around their star, just as the planets of the Solar System do. So the Solar System, with its numerous planets, may not be so unusual, and as astronomical techniques improve, scientists may someday be able to observe and explore planets like Earth orbiting other stars.

How do planets form? Historically, there have been two types of explanations. One type, first advanced by Compte de Buffon in 1745, accounts for the planets by interaction between the Sun and another star. According to these models, the planets were torn from the Sun by the gravitational attraction of a passing star. Such explanations, however, are unable to explain the existence of planets. Calculations show that material drawn from the Sun by the gravity of another star, in the unlikely event one should pass close enough, would either fall back into the Sun or be dispersed into space before it could coagulate into planets. A mechanism in which the Sun and planets formed together, as first proposed by René Descartes in 1644, seems much more likely.

In many ways, the early history of a star is analogous to that of a galaxy. Irregularities cause the interstellar gas and dust to fragment into smaller clouds. In a few of these clouds the conditions are sufficient to cause the cloud to collapse and a primitive star nebula to be born. Invariably, turbulence causes the nebula to rotate and flatten into a disk. If the density of matter at the center of the nebula is great enough, then the nuclear reactions required for a new star will begin. The Sun had such a beginning.

The Birth of Earth

Within the primitive Solar Nebula, dust and ices of water, ammonia, and methane collided and fused to form larger particles, perhaps as much as a centimeter or so in diameter. These tiny clumps of material settled through the gas of the nebula to the midplane of the flattening disk, where additional collisions and the increasing gravity of the growing clumps caused the particles to adhere to one another and to form larger bodies. Some of these bodies, called planetesimals, of which there may have been hundreds of thousands, reached diameters of several hundred kilometers. The accumulation of heat generated by the infall and collision of the aggregating particles and by the decay of short-lived radioactive elements caused the interiors of the larger planetesimals to reach temperatures sufficient for the rocks to partially melt. This resulted in the formation of new rocks, in volcanic eruptions on the surface, and in the formation of nickel-iron cores.

Temperature differences within the solar nebula caused considerable separation of matter, with the "rocky" materials being concentrated nearer the Sun and the bulk of the gaseous material gathering farther away. This accounts for the clustering of the higher-density terrestrial, or rocky, planets—Mercury, Venus, Earth, and Mars—close to the Sun, and the lower-density gaseous and icy planets—Jupiter, Saturn, Uranus, Neptune, and Pluto—in more distant orbits (see Figure 2.2).

The final stages of the formation of the Solar System involved the aggregation of planetesimals into proto-planets that were only a fraction of the size of the planets they would eventually become. Calculations show that the path from pre–Solar Nebula to proto-planets took only about 5 million years. As the proto-planets grew, they became increasingly efficient at attracting and holding stray bodies, with the result that early in their history they swept their orbits free of extraneous matter, growing to their present sizes in the process. It probably took only 50–100 million years for Earth to reach its present size, complete with metallic core and orbiting Moon. Evidence of the vigorous cleaning process in the Solar System, which resulted in a heavy bombardment of the planetary surfaces, is visible in the giant craters still visible on the Moon, Mars, and Mercury. This bombardment with large impactors continued for the first 700 million years of Earth's history. One such collision of a body about the size of Mars with the proto-Earth only about half its present size resulted in the formation of the Moon. Most of the gas and small dust particles not captured by the Sun and larger planets were blown out of the Solar System by radiation, in the form of light and fragments of atoms, from the Sun. Although the "cleaning" of the Solar System was rather efficient, it was not totally so. The billions of comets that reside mostly in orbits beyond Pluto, the meteoroids, and the asteroids represent the uncollected debris of the solar nebula.

Even before Earth had reached its present size, the energy of the numerous impacts involved as it collected material, and the heat generated by the short-lived radioactive elements, had resulted in melting of the outer several hundred kilometers. A thin crust may have covered this magma ocean, but if so, the crust would have been in a constant state of creation and destruction. Within the first 10 million years, the surface of Earth began to cool and lose gases from its interior, forming an atmosphere and oceans, but neither was exactly like today's. The early atmosphere consisted primarily of nitrogen and carbon dioxide with some water vapor, methane, and ammonia. Oxygen would come nearly 2 billion years later, generated by single-celled photosynthetic organisms. The atmospheric carbon dioxide, perhaps 100 times more abundant than today, caused the surface of Earth to be hot because of the greenhouse effect. The waters of the shallow oceans were rich in sulfuric and hydrochloric acid, and they vigorously attacked the primordial rocks being erupted from the numerous volcanoes.

Early Earth would not have been a nice place to vacation. But chemistry can work miracles. Within 10 million years or so, the interaction between the atmosphere, the oceans, and the lava flows had resulted in the transfer of most of the carbon dioxide in the atmosphere to the ocean and the ocean floor. The surface of Earth cooled and the acids in the ocean were neutralized, setting the stage for the eventual appearance of the first anaerobic cyanobacteria about 3.5 billion years ago. But the origin of life on Earth is another story, not told here.

As much as scientists would like to, it is impossible to determine the age of Earth directly by measurements on rocks and minerals. This is because Earth is a very active planet, and its turbulent history has effectively erased almost the entire geologic record of the first half-billion years. The rocks formed during that period of Earth's history have not been found and may no longer exist. The intense heat, bombardment, and chemical activity resulted in the melting of old rocks and the formation of new ones. The heat also initiated convection, the ponderous plastic flow of rocks driven by the combination of gravity and temperature differences, which began to redistribute and recycle the rocks within the Earth.

These processes, which continue today, soon resulted in a heavier core of iron and nickel, a thin crust of lighter rocks, and a mantle, the region between the core and the crust that consists of rocks of intermediate density. Furthermore, mantle convection is an ongoing process whereby new crust is continually created and recycled, continents collide and are torn apart, rocks are folded, faulted, uplifted, weathered, eroded, buried, and changed chemically and physically. As a result of these relentless processes, none of Earth's earliest-formed rocks have survived—at least none have yet been found. Radioactive dating techniques show that the oldest Earth rocks found so far are 4.0 billion years old; the oldest minerals are 4.4 billion years old. While these ages provide some valuable information about Earth's early history, they provide only minimum ages for the planet.

The best evidence for the age of Earth does not come primarily from Earth at all, but from meteorites. Fortunately, the formation of the Solar System, of which Earth, the Moon, and meteorites are an integral part, occurred over a very short period of time, probably about 50–100 million years. Thus, the lack of direct evidence from Earth itself does not introduce a very large error in the age of Earth.

Evidence for the ages of meteorites, the Moon, and Earth is discussed in more detail in the chapters that follow. First, however, it is important to examine a few of the many historical attempts to determine Earth's age. These early attempts are interesting in their own right, and they nicely illustrate the nature and evolution of the centuries-long struggle that scientists endured in their attempt to resolve the question of Earth's age. They also illustrate the types of dating methods that do not work.

CHAPTER THREE

Early Attempts

When was the first attempt to assign an age to Earth, and by whom? We will probably never know, but it is clear that humans have been seeking an answer to this interesting question for more than two millennia (Table 3.1). For most of this time the question was in the hands of theologians, who attempted to provide answers based on theological theory and sacred writings. These religion-based ages for Earth ranged from a few thousand years to infinity, but those derived from the Old Testament were invariably in the range of 5–10 thousand years. It was not until the eighteenth century that the first pioneering naturalists began to think of methods for estimating Earth's age based on observations, measurements, and scientific principles.

During the latter half of the nineteenth century, the age of Earth was one of the most hotly debated subjects of science. On the one hand were those scientists, largely physicists, who sought precise answers from rigorous calculations, even though their methods were based on questionable assumptions and their data were poor. Among these methods, the best known were the cooling of Earth and the accumulation of salt (actually the element sodium) in the ocean. The results of such calculations varied widely, but they were usually on the order of a few tens of millions to a few hundred million years.

On the other hand were those naturalists, primarily geologists and biologists, who were just beginning to look closely at the rocks and living things of Earth and ponder the ways in which they might have formed. In nature's thick accumulation of sedimentary rocks and the fossils they contained, these naturalists saw the products of processes operating over vast periods of time. From the observed rates of the same processes in the present world they could estimate, though not then prove, that billions of years were required for Earth's history. Ironically, the geologists and biologists were correct, and the physicists would eventually provide the tools to prove them so.

In general, the values assigned by scientists to the age of Earth have increased several orders of magnitude over the past century or so. Why such an enormous change? There are two reasons. The first reason is the ever-increasing understanding of Earth and of the nature and rates of geologic processes over time. This understanding came about because of the acquisition of new instruments and methods with which to investigate and understand Earth and its physical place in the cosmos. In short, it is the result of scientific progress—the birth and maturation of geology and related sciences. The second reason is that the increase over time in the values for the age of Earth is a consequence of the gradual separation of scientific and theological thinking about nature.

Prior to the mid-eighteenth century the Christian Church dominated Western thinking not only in affairs of the soul, but also in matters of nature. The book of Genesis, as translated and interpreted by religious scholars, was the principal and final authority for understanding both the relationship of man to God and the physical history of the Universe. Taken literally, the biblical measures of time were the day and the generation—time reckoned in millions and billions of years was unthinkable because Scripture, it was then thought, proved otherwise. It was not until the Age of Enlightenment in the seventeenth and eighteenth centuries that observation and secular reasoning became widely accepted as an alternative to the Christian interpretation of nature, and modern science was born. But the habits of centuries were difficult to overcome. The process was slow and required fundamental and difficult changes in both scientific and theological thought.

Estimates of the age of Earth prior to about 1950 are all wrong because they are based on methods now known to be invalid. Nevertheless, they are important not only for historical reasons, but also because they provide an interesting and valuable perspective on the current understanding of the age of Earth. It is, therefore, well worth describing some of the more interesting and important of these estimates and the methodology and people who supported them.

Table 3.1

Some Early (pre–1950) Estimates of the Age of Earth

Basis	Author	Year	Age of Earth
RELIGION			*Years*
Hindu chronology	Priesthood	120–150 B.C.	1,972,949,081
Biblical chronology	Theophilus of Antioch	169	7,529
Biblical chronology	St. Basil the Great	4th century	6,004
Biblical chronology	St. Augustine	5th century	6,331
Biblical chronology	James Ussher	1650	6,004
Movement of solar apogee	Johannes Kepler	ca. 1620	5,993
SEA LEVEL			*Million Years*
Decline of sea level	Benoit de Maillet	1748	>2,000
TEMPERATURE			
Cooling of Earth	Comte de Buffon	1774	0.075
Cooling of Earth	Lord Kelvin	1862	20–400
Cooling of Earth	P. G. Tait	1869	10–15
Cooling of Earth	C. King	1893	24
Cooling of Earth	Lord Kelvin	1897	20–40
Cooling of Sun	H. L. F. von Helmholtz	1856	22
Cooling of Sun	Lord Kelvin	1862	10–500
Cooling of Sun	S. Newcomb	1892	18
ORBITAL PHYSICS			
Earth-Moon tidal retardation	G. Darwin	1898	>56
Earth tidal effects	P. G. Tait	1876	<10
Earth tidal effects	Lord Kelvin	1897	<1,000
Change in eccentricity of Mercury's orbit	H. Jeffreys	1918	3,000
OCEAN CHEMISTRY			
Sulfate accumulation	T. M. Reade	1876	25
Sodium accumulation	J. Joly	1899	89
Sodium accumulation	J. Joly	1900	90–100
Sodium accumulation	J. Joly	1909	<150
Sodium accumulation	W. J. Sollas	1909	80–150
Sodium accumulation	G. F. Becker	1910	50–70
Sodium accumulation	A. Knopf	1931	>100

Table 3.1 (continued)

Basis	Author	Year	Age of Earth
EROSION AND SEDIMENTATION			
Limestone accumulation	T. M. Reade	1879	600
Limestone accumulation	A. Holmes	1913	320
Sediment accumulation	A. Geikie	1868	100
Sediment accumulation	T. H. Huxley	1869	100
Sediment accumulation	S. Haughton	1871	1,526
Sediment accumulation	A. Winchell	1883	3
Sediment accumulation	W. J. McGee	1892	15,000
Sediment accumulation	C. D. Walcott	1893	35–80
Sediment accumulation	W. Upham	1893	<100
Sediment accumulation	J. Joly	1908	80
Sediment accumulation	W. J. Sollas	1909	80
Sediment accumulation	J. Barrell	1917	1,250–1,700
RADIOACTIVITY			
Decay of U to Pb in crust	H. N. Russell	1921	2,000–8,000
Decay of U to Pb in crust	A. Holmes	1927	1,600–3,000
Decay of U to Pb in crust	E. Rutherford	1929	3,400
Decay of U to Pb in minerals	A. Knopf	1931	>2,000
Pb isotopes in Earth	E. K. Gerling	1942	3,940
Pb isotopes in Earth	A. Holmes	1946	3,000
Pb isotopes in Earth	H. Jeffreys	1948	1,340
Decay of Rb isotope to Sr isotope	A. K. Brewer	1938	<15,000
Abundances of radioactive isotopes	H. E. Suess	1949	4,000–5,000

NOTES: Not all are ages for Earth. Some are for very early events in Earth's history, such as the creation of man, the age of the oceans, etc., while others are for the age of the Solar System or the age of matter. Ages of less than 10,000 years have been corrected from the year of publication to the year 2000 where appropriate. None of these methods gives the correct age of Earth.

SOURCE: From a compilation by Dalrymple 1991.

de Maillet and the Decline of the Sea

One of the earliest departures from the Christian tradition of a very young Earth using observation and inductive reasoning was by Benoit de Maillet (1656–1738), a French diplomat, savant, and amateur naturalist. From age 35 until his retirement in 1720, he held various diplomatic posts around the Mediterranean, which not only led him to travel extensively but also afforded him the opportunity to observe and study the geology and geography of the area.

The result of de Maillet's extensive studies and observations was a history of Earth that required a span of time vastly greater than the few thousand years calculated by biblical chronologists. Because he was well aware of the power and influence of the Church, de Maillet explained his ideas in a fictitious account that was not published until 10 years after his death. He told of a series of conversations between a French missionary and an Indian philosopher named Telliamed (de Maillet spelled backwards). Speaking through Telliamed, de Maillet revealed a history of Earth that attempted to account for its origin, the formation of sedimentary rocks and fossils, the building of mountains, and the origin of all forms of life, including man.

De Maillet accepted the idea of René Descartes (1596–1650), that Earth was once covered entirely with water. This was not simply an assumption; it was based, at least partly, on observations. Telliamed reveals to the missionary that his grandfather had found that the layered rocks of the mountains far inland contained sea shells, certain proof of formation over a long period of time in an ocean more extensive than the present one. Telliamed observes that the water that once covered the globe is still being lost into the "vortex" of the Solar System, causing sea level to fall continuously. It is this decline of sea level that is the basis for de Maillet's estimate of the age of Earth.

Telliamed's grandfather had observed that part of the shoreline near his village had been awash when he was a boy and had emerged some years later. He reasoned that he could calculate the approximate time that the tallest mountains first emerged from the sea if he knew the rate at which sea level was declining. To this end, his grandfather constructed a hydrographic station on the shore near his village and over a period of 75 years he determined that the decline amounted to 3 inches per century. To confirm these observations, Telliamed cites structures at Carthage, Alexandria, and Acre that were originally constructed at sea level but now stand several feet above. Telliamed asserts that certain ancient settlements, some now standing as much as 6000 feet above sea level, originated as seaports. Using the rate of 3 inches per century, the age of these oldest marine settlements must be 2,400,000 years or so. Telliamed concludes that on any reasonable assumption, 2 billion years must have passed since the decline of sea level began.

As strange as they may sound today, de Maillet's ideas were an honest and pioneering attempt to interpret Earth history on the bases of observations and the scientific theories of his time. It is now certain that Earth was never covered entirely by water. Most of the changes in sea level cited by him as evidence that the sea was declining are due to local uplift of the land, and he failed to recognize that there are also areas around the Mediterranean that are sinking. Thus, de Maillet's basis for calculating the age of Earth from declining sea level is invalid. Nevertheless, he was one of the first to recognize the importance of the slow operation of natural processes over vast periods of time in forming Earth's rocks and shaping its features. He also recognized that the timing of events in the history of Earth could be estimated by observing natural processes, measuring their rates, and making reasonable and logical calculations. In one forceful intellectual blow, he felled the notion that geologic time must be reckoned in terms of the human lifetime or be based on biblical chronology, and he introduced the concept of an Earth that is billions, instead of mere thousands, of years old.

The Cooling of Earth

In 1862, Lord Kelvin (William Thomson) published his first calculation of the age of Earth based on the time required for the planet to cool from a white-hot, molten globe to its present state. For more than half a century, Kelvin's ideas dominated scientific thinking about the antiquity of Earth. The idea that a date for Earth's beginning could be found from cooling calculations, however, did not originate with Kelvin but had its inception nearly two centuries before.

Isaac Newton (1642–1727), based on experiments on heated bodies cooling in air, calculated that a globe of hot iron the size of Earth would require more than 50,000 years to cool, but he did not suggest that this was Earth's age. Baron Gottfried Wilhelm von Leibniz (1646–1716) was an early subscriber to the concept of an initially molten Earth. The cooling Earth, according to Leibniz, was sculpted by large bubbles, some of which hardened into mountains while others collapsed to form valleys. Like Newton, Leibniz did not venture to determine an age for Earth from cooling. That bold step was left to another prominent figure of the Enlightenment.

Georges-Louis Leclerc (1707–1788), Comte de Buffon, was one of the most productive and well-known scientists of the eighteenth century. He is best known for an encyclopedic work in which he attempted to synthesize all knowledge of nature and natural history into an intelligible whole. *Histoire Naturelle, Générale et Particulière* was originally intended to include an ambitious fifty volumes, of which Buffon actually completed thirty-five before his death, including *Epochs of Nature.* In it Buffon divided the history of Earth into seven epochs. In the first epoch, Earth was a molten globe. The final epoch included the advent of man and the world as it is today. The intervening epochs included the formation of Earth's surface, the appearance of oceans and the beginning of life, the formation of continents, the development of mammals, and the separation of the American and Eurasian continents.

According to Buffon, the first epoch began when a comet collided with the Sun, causing the ejection of hot gases and liquid to form the planets and moons of the Solar System. Buffon's next step was to calculate the time required for Earth to cool from its initial molten state to its present state. Rather than speculate, Buffon had his foundry fabricate ten iron spheres whose diameters varied in half-inch increments up to 5 inches. He heated them until they were white hot and then observed the time required for them to cool. He found an approximately constant relationship between diameter and cooling time, which he then logically but naively extrapolated to a sphere the size of Earth. On this basis he calculated that a mass of molten iron the size of Earth would require 96,670 years to cool to the present temperature of Earth.

Buffon performed similar experiments on a second set of spheres made of materials nearer the actual composition of Earth. He corrected his calculations for the heat added to Earth by the Sun and combined these data with some major events in Earth's history, as reconstructed in *Epochs*, to deduce the following scale of times, each in years from the beginning:

	Years
Surface of Earth consolidated	1
Earth consolidated to its center	2,936
Earth cool enough to be touched	34,270
Beginning of life	35,983
Temperature of present reached	74,832
End of life	168,123

Although the calculations were detailed, Buffon was suspicious of his results because he felt that the events of Earth history required a much longer time than the 75,000 years his calculations indicated, perhaps as much as double. He was impressed, for example, by the tremendous thickness of sedimentary rocks exposed in the Alps and by the exceedingly slow rate at which similar sediments are formed in the modern ocean. In unpublished manuscripts not publicized until the century after his death, Buffon detailed several longer chronologies, including one that estimated the age of Earth at nearly 3 billion years.

There are numerous complicating factors, described below, that preclude calculating an age for Earth based on cooling. Most of these complications could not have been anticipated by Buffon given the state of scientific knowledge in the eighteenth century. Despite the fact that Buffon's results were wrong, he was the first to apply laboratory experimentation to the problem of the age of Earth, and in doing so became one of the founders of geophysics. Nearly a century was to pass, however, before Buffon's approach was explored in detail.

Two papers appeared in 1862, written by a professor of the university at Glasgow, that would dominate scientific thinking about the age of Earth for more than five decades. The papers concerned the cooling of the Sun and Earth, and their author was William Thomson (1824–1907), who would later become Lord Kelvin.

Kelvin's interest in the age of Earth arose from his research in thermodynamics, which is the physics of heat and its relation to other forms of energy. One consequence of the second law of thermodynamics is that natural engines, such as Earth, the Sun, and the Solar System, must eventually run down—they will cease to function when all of their heat is gone. To Kelvin this meant that both Earth and the Sun must be cooling. In his paper entitled "On the Secular Cooling of the Earth," read to the Royal Society of Edinburgh in 1862 and published in their *Transactions* 2 years later, Kelvin argued that the energy of the entire Solar System must suffer "irrevocable loss," and that given sufficient data the age of Earth could be determined from the present distributions of temperatures within Earth's crust.

The basis for Kelvin's calculations was relatively simple. Any uniform body heated to high temperature and then placed in colder surroundings will lose heat to those surroundings in a predictable way. The surface of the body will cool to the surrounding temperature very quickly and remain there. The center of the body will remain at the initial temperature for a longer time. The temperatures in the interior of the body thus will increase from the surface inward, and a plot of temperature as a function of depth will be a curve—the geothermal gradient—that depends on four quantities: the initial temperature, the surface temperature, the thermal conductivity of the body (the rate at which it conducts heat), and time. From the initial temperature, the geothermal gradient, and the thermal conductivity of Earth, therefore, Kelvin could calculate how long Earth had been cooling; this number was equivalent to its age.

Although the existing temperature data were far from the complete survey that Kelvin desired, there were enough measurements from mines and wells to show that the temperature in Earth increased with depth and, therefore, that heat was flowing outward from Earth's interior. To Kelvin, these data strongly suggested that Earth was indeed cooling. He admitted that some heat might be generated by the tidal effects of the Moon and Sun or by chemical action, but on the whole he thought that these sources were inadequate to account for any but a small fraction of the heat flowing from Earth.

Kelvin needed to make some assumptions in order to proceed with his calculations. Among them were that Earth began at some uniform high temperature. Another was that all of the heat was transferred from the interior to the surface by conduction. Another was that Earth was homogeneous. Yet another was that Earth was flat and infinite in all directions. Even though the Earth is very nearly a sphere, Kelvin argued that the errors introduced by assuming a flat Earth were negligible, and the assumption made the calculations a lot easier.

The data available to Kelvin were not very good. As the initial temperature Kelvin used 7000° F (3871° C), which was then thought to be a reasonable estimate for the temperature at which rock melts. At the time, however, there were no experimental measurements, and 7000° F is actually three to four times too high. For the thermal conductivity, Kelvin used the average of three measured values on sand, sandstone, and "trap rock" (a rock similar in composition to the basalt lava found in Hawaii) that he had determined in the laboratory and reported to the Royal Society of Edinburgh in 1860. There were a number of measurements of geothermal gradients available from wells and mines, but they varied widely and were insufficient to provide any meaningful worldwide average. Kelvin, therefore, adopted the commonly accepted value of 1° F for each 50 feet of depth. Using these data, Kelvin calculated that Earth solidified 98 million years ago.

Kelvin knew that the data then available were inadequate, so he allowed for errors by broadening the limits for the time of solidification:

> But I think we may with much probability say that the consolidation cannot have taken place less than 20,000,000 years ago, or we should have more underground heat than we actually have, nor more than 400,000,000 years ago, or we should not have so much as the least observed underground increment of temperature. (Thomson 1864, p. 474)

On a similar basis, and assuming that the Sun's fires were kept ablaze by the gravitational energy released by the continuing infall of meteoritic rocks and dust, Kelvin calculated that the Sun's age was probably less than 100 million years and certainly less than 500 million years.

In spite of the questionable assumptions and the high degree of uncertainty in the data, Kelvin's calculations of the ages of the Sun and Earth were, at the time, considered highly authoritative. For three decades they remained as the best that physics could offer on the subject. Nineteenth-century geologists had finally broken with the Church-inspired doctrine of a recent creation and were formulating new concepts for the history of Earth and its living things based on the availability of seemingly limitless time. Among these geologists was Charles Darwin (1809–1882), whose remarkable new theory of descent with modification, now known as the theory of evolution, required that geologic time be long enough to account for the origin of the numerous species of plants and animals, both living and fossil, through a slow process of natural selection. Now Kelvin, one of the most imposing scientific authorities of the day, had calculated rather more narrow limits on the time available to geologists and biologists—limits that were derived from data, physical principles, and elegant mathematics. The limits on the age of Earth from cooling calculations, however, were to get narrower still, and a noted geologist would provide the data, the rationale, and the calculations.

Clarence King (1842–1901) was a flamboyant, well-known public figure. From 1868 to 1878, he led the Geological Survey Along the Fortieth Parallel from 1868 to 1878. This gave him considerable public recognition, and in 1879 he was appointed the first Director of the new U.S. Geological Survey (USGS). Among his accomplishments, he discovered the first glaciers in the United States, discovered Mt. Whitney in the Sierra Nevada, and publically exposed a fradulent Colorado diamond field. Recognizing the potential importance of the application of physics to the study of Earth, King established a program of geophysics within the USGS, hiring Carl Barus, a physicist from Würzburg, Germany.

Even during King's time, it was known that the melting temperature of a solid increases with increasing pressure. At King's behest, and partly supported by King's personal funds, Barus set about determining the melting and solidification temperatures as a function of pressure for diabase, an igneous rock chemically similar to the basalt lava that forms the Hawaiian Islands. Pressure apparatus to make these measurements directly was unavailable then, so Barus did the next best thing—he measured the changes in volume and the heat released as the diabase cooled and solidified from a melt at one atmosphere. He then used these data in a well-known equation (inapplicable to rocks) to calculate the change in melting temperature caused by the increase in pressure with depth below Earth's surface. King now had the information he needed.

He approached the problem of a cooling Earth from a very simple but elegant point of view. Kelvin and other workers had previously argued that Earth was effectively rigid because the tides in the solid Earth caused by the Moon and Sun would be very much greater if Earth were liquid. King used Kelvin's conclusion to test the validity of various models for the change of temperature with depth below Earth's surface. Although an initially molten Earth was acceptable, any increase in temperature with depth that resulted in a present-day layer of fluid rock in the outer one-quarter of the Earth's radius could be excluded from further consideration; if such a liquid layer existed, the tides would cause violent disruptions of the overlying solid crust.

King's first step was to extend Barus's value for the change in melting temperature as a function of pressure from the surface clear to the center of the Earth. This was a bold step on King's part, for Barus's measurements were all made at a pressure of 1 atmosphere and at temperatures less than 1500° C, and such an extrapolation did not take into account the possibility of changes in either physical properties or rock type with depth. One important consequence of Barus's new data and King's calculations was the revelation that Earth was probably never entirely molten. Because the melting temperature of diabase increased with pressure, and therefore with depth, unreasonably high initial temperatures were required for a completely molten Earth. For example, if King's extrapolation was correct, an initial temperature of 76,000° C was required to create liquid diabase at Earth's center. King rejected such temperatures as entirely unreasonable and decided that the initial temperature of Earth must have been such that the planet did not begin its life as a liquid ball. On this score he was right.

Having established the melting conditions for a diabase Earth, King then calculated the change in temperature with depth for various initial Earth temperatures and cooling times using the same assumptions, mathematics, and thermal conductivity as Kelvin had used in his 1862 paper. Finally, he tested the validity of these temperature gradients by seeing which of them resulted in a solid Earth, with a combination of initial temperature and cooling time (age of Earth) that was only slightly greater than required to do the job. King settled on an initial temperature of about 2000° C and on 24 million years as the age of Earth.

Lord Kelvin's last word on the age of Earth from cooling calculations was in 1897, but he added little new to his previous methods or results. In an address to the Victoria Institute entitled "The Age of the Earth as an Abode Fitted for Life," he noted that critical data lacking in 1862 had become available, and that these new data greatly modified the results of his previous calculations. Although he did not elaborate on exactly what he had done, his new result, he said, supported that of King.

The conclusions of King and Kelvin regarding the probability of a youthful Earth were widely accepted, but not all scientists were convinced. In 1869, for example, Thomas H. Huxley attacked the assumptions and data used by Kelvin in his first calculations, pointing out the high degree of uncertainty in both. Moreover, John Perry, a noted physicist and former assistant of Kelvin's, questioned the assumptions used by Kelvin and King and showed that cooling calculations using different but equally likely assumptions and data resulted in an age for the planet of as much as 29 billion years.

One of the harshest and most carefully measured criticisms of Kelvin's results came from Thomas C. Chamberlin (1843–1928), a highly respected professor of geology at the University of Chicago, who was skilled in physics and mathematics. In a paper specifically directed at Kelvin's 1897 address, Chamberlin challenged the noted physicist on his own ground. He noted that Kelvin had greatly overstated the certainty of his assumptions and results, and that Kelvin's use of such phrases as "no other possible alternative" and "certain truth" were unjustified. In particular, Chamberlin challenged Kelvin's "very sure assumption" that the Earth was once a white-hot liquid.

Chamberlin argued that the slow formation of Earth from an accumulation of the rocks and dust of the Solar System was far more probable than the rapid accumulation assumed by Kelvin. Slow formation would provide the internal heat necessary to explain the measured geothermal gradient, but would not result in a liquid planet. Chamberlin also showed that the rapid accumulation assumed by Kelvin would result in an initial Earth temperature exceeding 3000° C, a condition that was precluded by the new data on the temperature and pressure of rock melting. In contrast, slow accumulation would result in a continued release of heat over a long period of time.

Chamberlin also attacked Kelvin's age for the Sun, pointing out that the existing knowledge of the history and evolution of the Sun was so meager that there was no valid way to determine the amount of energy available or the timing of its release. He also argued that the behavior of matter in the interior of the Sun was completely unknown, and that there might be sources of energy in addition to those considered by Kelvin. In particular, he argued that atoms might be "complex organizations and the seats of enormous energies" (Chamberlin 1899, p. 225), which could prolong the Sun's fires far beyond the limits set by Kelvin. This conjecture was truly prophetic, for although the phenomenon of radioactivity had been discovered by Henri Becquerel 3 years earlier, virtually nothing was known about the nature of atoms or the tremendous energies they contain.

The question of whether or not calculations based on simple cooling could provide valid estimates for the ages of the Sun and Earth was partly resolved in 1903, when Ernest Rutherford and Frederick Soddy first determined the amount of heat generated by radioactive decay and noted that their discovery had an important bearing on cosmological hypotheses: "The maintenance of solar energy, for example, no longer presents any fundamental difficulty if the internal energy of the component elements is considered to be available . . ." (Rutherford and Soddy 1903, p. 591). It is now known, of course, that the Sun's fires are due to nuclear reactions and that its fuel will last for another 5 billion years or so.

Although Kelvin, King, and their followers claimed a high degree of accuracy for their cooling calculations, their results were, as Chamberlin and others had shown, subject to large errors. By carefully choosing from a wide range of reasonable values for the geothermal gradient, initial temperature, and thermal conductivity, it is possible to calculate ages for Earth from simple, conductive cooling models that exceed 3 billion years.

But the most serious flaw in Kelvin's method is that the basic assumptions are wrong. Earth's heat production and loss are far more complex than Kelvin, King, Chamberlin, or any nineteenth-century scientist could possibly have imagined. One complication is that there are far more sources of heat within Earth than were known in Kelvin's time, including heat from the gravitational energy released as the growing Earth collected material from space, from the decay of radioactive elements, and from the heat released as the liquid core crystallizes. Another factor that invalidates Kelvin's approach is that convection in the solid mantle—that is, the slow heat- and gravity-driven flow of Earth's deep rocks—is a far more important mechanism for the loss of heat from within Earth than conduction. Finally, there are other poorly known factors, including the exact composition and structure of the deep Earth, as well as the exact physical properties of the rocks at depth. Thus, heat production and heat loss in planet Earth are very complicated, and current knowledge of the relevant details is far too inadequate to permit any valid estimates of the age of Earth from thermal calculations.

Darwin's Origin and Age of the Moon

George Darwin (1845–1912) was the second son of Charles Darwin, the renowned British naturalist. Although now overshadowed by his famous father, George Darwin pioneered in the study of tides and the origin of the Universe, and he is considered by many to be the father of modern geophysics. During his day, he was a highly respected scientist and was eventually knighted by King Edward VII for his accomplishments.

George Darwin is probably best remembered as the originator of the hypothesis that the Moon was torn from Earth by rapid rotation. This idea was far from being one of his most significant accomplishments, and he gave the idea little credibility, as do modern scientists. Nonetheless, his calculations provided an early and widely cited estimate for the age of Earth.

The basis of Darwin's idea was the braking effect of the tides on the rotation of Earth. The gravitational pull of the Moon (and Sun, but the Moon's effect is much greater) causes tides in the solid Earth, as well as in the oceans. As these tides move around Earth, energy is lost in the form of heat, which is radiated into space. This loss of energy is a kind of "drag," or friction, that results in a slowing of Earth's rotation. Although this phenomenon was discovered by Immanual Kant in 1754, it was not until 1868 that Lord Kelvin first used tidal friction and the resultant slowing of the rate of rotation of the Earth to calculate an upper limit for the age of Earth of 1 billion years.

Darwin carried the concept that tidal friction might provide an estimate of Earth's age to a much higher degree of refinement than Kelvin had, taking into account the effects of the tides on both Earth and Moon. Darwin reasoned that tidal friction must not only be causing the planet's rotation to slow, resulting in a lengthening of the day, but it must also be increasing the time it takes the Moon to orbit Earth and the distance from Earth to Moon. Darwin used the analogy of a rock twirled at the end of an elastic string—the more vigorously it is twirled, the more the string stretches and the longer it takes to make a complete 360-degree circuit (orbit).

Darwin first calculated what would happen in the future. He showed that when the lengths of both the day and the month equal 55 of our present days, then the Earth-Moon system will be stable, and Earth and Moon will forever present the same side toward each other as if connected by a rigid bar. Next he attempted to determine what happens when this line of reasoning is followed back in time. He found that both the day and the month shorten until once again the configuration reaches stability. Darwin calculated that this occurs when both the day and the month are about 3 to 5 of our present hours in length, at which time the two bodies again revolve as if connected by a rigid bar, and the Moon is only about three Earth radii from Earth.

Up to this point, Darwin's calculations were fairly straightforward and reasonable. To go further, however, he was forced to engage in speculation, for his mathematical analysis was incapable of exceeding the period of initial stability. He suggested that an original molten Earth rotating rapidly might have spun off large pieces that aggregated to form the Moon, and he calculated that an initial Earth rotation of 1.5–2 hours would do the job. Once the Moon has been ejected from Earth, the path to the present is not overly difficult. To overcome the stability of the 3-to-5-hour period of rotation, Darwin suggested that the Moon must originally have circled Earth slightly slower than Earth rotated on its axis. Tidal friction would then take over, causing both the rotation of Earth and the orbital period of the Moon to lengthen, and the orbital distance to increase irreversibly.

Almost as an afterthought, Darwin calculated the length of time required for Earth and Moon to progress from the initial conditions, a 3-to-5-hour day and month, to the present with a day of 24 hours and a lunar month of 27.5 days. He found that 56 million years would be the minimum length of time required. Darwin was not strongly attached to the validity of his hypothesis and considered it to be purely speculative.

We now know that Darwin's minimum value for the age of Earth is much too low and his origin of the Moon is incorrect. Among other serious difficulties, the effects of tidal action can be changed materially by even moderate changes in the shapes of the oceans and continents, such as have occurred over geologic time, so extrapolation of present rates into the past is likely to result in serious errors. Within the framework of late nineteenth-century science Darwin's hypothesis was reasonable and appropriate, but we now know that tidal considerations are incapable of providing a basis for determining Earth's age. I suspect Sir George wouldn't have minded.

The Salt Clock

Imagine a tub of water to which a chemical is continuously and constantly added. If you knew the amounts of the chemical in the water both now and when the tub was first filled, and the rate at which the chemical was added to the water, you could calculate the time at which the addition of the chemical began. It was the possibility of just such a calculation that Edmund Halley (1656–1742), the Astronomer Royal who predicted the return of the comet that bears his name, had in mind in 1715 when he proposed that the age of Earth might be calculated from the salt content of the oceans and of certain kinds of lakes.

Halley observed that all lakes that receive runoff from rivers but lack outflow contain salt in varying amounts. The concentration of salt in the waters of these lakes must increase, he said, because salt, picked up by the rivers in their passage over Earth, is continuously added but not removed. If this was the actual cause of the saltiness of the lakes, Halley reasoned, then it is probable that the same mechanism was responsible for the saltiness of the oceans.

Halley reasoned that if the concentration of salt in the oceans was measured at different times, then the rate of addition and the age of the oceans, which he equated with the age of Earth, could be determined. Unfortunately, the data to make such a calculation were lacking, and Halley could only lament that the ancient Greek and Latin authors had not provided information on the salinity of the oceans 2000 years ago.

Halley's idea was largely forgotten until 1876, when T. Mellard Reade (1832–1909), a geologic and civil engineer, rediscovered the method he called *chemical denudation*, a name he choose to describe the removal of elements from the continents and their deposition in the oceans. Reade proposed that the age of the oceans could be found using the concentrations of certain salts—namely, those of chlorine (known as chlorides) and sulphur (sulphates). Instead of determining the rates of addition of these compounds by measuring the concentrations at different times, as Halley had proposed, Reade attempted to estimate the annual amounts carried into the oceans by the rivers of the world. At the present annual rates of addition, he calculated that it would require 25 million years for the sulfates of calcium and magnesium to reach their present concentrations in ocean water; for chlorides (principally of sodium), the comparable time was 200 million years.

Reade's basic idea of dating Earth from the progressive change in the chemistry of the oceans was refined by John Joly (1857–1933), a professor of geology and mineralogy at the University of Dublin. Joly's classic paper, "An Estimate of the Geological Age of the Earth," was read to the Royal Dublin Society in 1899. Joly proposed to measure the age of Earth from the accumulation of the element sodium, which, along with chlorine, makes common salt (sodium chloride). Joly's basic approach was the height of simplicity—the age of Earth was equal to the total sodium in the oceans divided by the amount of sodium added to the oceans from the world's rivers in a year. This assumes, of course, that there was no sodium in the oceans to begin with.

Determining a value for sodium in the modern oceans was not particularly difficult. Sir John Murray, a British oceanographer, had made estimates of the mass and mean depth of the oceans, the total volume of river discharge, and the quantity of dissolved matter in a number of the world's rivers.

Joly used these data, along with the salinity of the oceans, to calculate the total mass of sodium in the oceans.

Murray had also provided estimates for the total annual river discharge into the oceans and of the total salts in river water. Joly used these values in his equation to get an uncorrected age for Earth of 99.4 million years. After correcting this age for the amount of sodium in the original oceans and the amount of sodium in the salt that was recycled by evaporation from the oceans and returned to the rivers in rain, Joly's final age for Earth was 89.3 million years. Joly considered other possible sources of error, including the sodium permanently removed from the system as salt deposits, and the possible violations of his assumption of a constant rate of sodium influx. Taken as a whole, however, he concluded they probably were insignificant.

The following year Joly revised his estimate upward to 90.8 million years and concluded that the probable age of Earth was 90–100 million years. This was not, however, his final word on the subject, and in 1909 he calculated an upper limit of 150 million years.

Joly was not the only one to play this game (see Table 3.1). In 1909, for example, the Oxford Professor William J. Sollas (1849–1936), using Joly's method, determined the age of the ocean to be between 80 and 150 million years.

George F. Becker, head of the Division of Chemical and Physical Research at the USGS, added a new dimension to the problem in 1910, when he proposed that the accumulation of sodium in the oceans was decreasing exponentially with time because the area of exposed crystalline rocks, the source of the sodium, was decreasing. He found that the age of Earth was between 70 and 50 million years, probably closer to 70 million.

What is wrong with the calculations of Reade, Joly, and those who adopted their methods? For one thing, the rates of erosion and solution, and the values of rainfall, runoff, continental area, and average exposed rock composition over geologic time, are so variable and poorly known that age calculations based on these quantities are mostly wishful thinking. But the most serious flaw in the method is in the basic assumption that elements and compounds accumulate continuously in the oceans. In fact, the composition of sea water is not changing significantly because it is in approximate chemical equilibrium. At present all elements are removed from the oceans at about the same rate at which they are introduced. In the case of sodium, slightly over half is recycled by evaporation and rain. The rest is removed as a constituent in sediments. The result of Joly's calculation was not an age for the oceans or Earth, but simply an estimate of the average time that sodium remains in sea water before it is removed. Chemical denudation was an important method for estimating the age of the oceans around the turn of the century. However, it is incapable of providing even a crude estimate of the age of Earth.

Sediment Accumulation

James Hutton (1726–1797), the physician-farmer who is generally considered the founder of modern geology, was the first to argue that the continents were formed from the ruins of pre-existing continents by processes similar to those acting today. In doing so, he provided the basis of a method that led to more estimates for the age of Earth than any other except biblical chronology. The method involved estimating the time required for the sedimentary rocks of Earth to accumulate.

Many nineteenth-century geologists were uncomfortable with the application of physics to the problems of the origin of Earth and the length of geologic time. A few, like Chamberlin and King, were not intimidated by the esoteric calculations of the physicists, but they were exceptions. Most preferred the conclusions drawn from the evidence of their own science. The use of sediment accumulation as an hourglass was a method based solely in geology. It was a game that any geologist could play and was considered by many to be the best weapon that geology had to offer in the heated debate with the physicists over the age of Earth.

The basic method was deceptively simple. It involved summing the thicknesses of sediment deposited during the various divisions of geologic time (Figure 3.1) and determining a rate for sediment deposition. The former divided by the latter was an estimate of the age of Earth, or at least the time the deposition of sediments first began in the world's oceans. There were many ways to do this and numerous uncertainties in the data and methods.

The scientist who most thoroughly applied the method was Charles D. Walcott (1850–1927). A noted American geologist renowned for his expertise in the stratigraphy (the study of stratified sedimentary rocks) and fossils of the Cambrian period, he was Director of the USGS from 1894 to 1907. Walcott was a cautious scientist, and rather than venture too far afield, he

Eon		Era/Subera		Period/Subperiod		Epoch	Age (million years ago)
Phanerozoic		Cenozoic		Quaternary		Holocene	0.01
						Pleistocene	1.6
			Tertiary	Neogene		Pliocene	5.2
						Miocene	23
				Paleogene		Oligocene	35
						Eocene	56
						Paleocene	65
		Mesozoic		Cretaceous			146
				Jurassic			208
				Triassic			248
		Paleozoic		Permian			290
				Carboniferous	Pennsylvanian		323
					Mississippian		362
				Devonian			408
				Silurian			439
				Ordovician			510
				Cambrian			570
Precambrian	Proterozoic						2500
	Archean						4000
	Priscoan						4490

Figure 3.1 The geologic time scale. Because of the lack of fossils, there is not nearly as much known about the Precambrian as there is about the Phanerozoic, and there is not universal agreement about how the Precambrian should be subdivided.

used the sedimentary rocks deposited in the Cordilleran Sea during the Paleozoic, a rock sequence that was the focus of much of his own research. The Cordilleran Sea was a shallow inland sea that once stretched from Mexico to Canada, and Walcott thought that the Paleozoic sedimentary rocks there had been deposited nearly without interruption.

Walcott divided the problem into two parts: (1) the time required for deposition of the detrital sedimentary rocks, such as sandstones and shales, which are composed of particulate debris; and (2) the time required to form the carbonate rocks, or limestones, which are formed primarily by chemical and organic precipitation from sea water. The detrital sedimentary rocks he divided into those formed during the early and middle Cambrian period, and those formed during the rest of the Paleozoic. He did so because the evidence indicated that the land area to the east of the Cordilleran Sea was depressed below sea level at the end of middle Cambrian time, thereby reducing the supply of sediment for the remainder of the Paleozoic.

Walcott's process was elaborate. For the early and middle Cambrian rocks he estimated their thickness, their area, the area of the land that supplied the sediment, and the rate of erosion and deposition, calculating a time for deposition of 0.50 million years. Using a similar method, his calculation of the time required to deposit the post-middle Cambrian detrital rocks yielded 0.66 million years.

The calculations involving the Paleozoic limestones were somewhat more complex than those for the detrital rocks. Complicated corrections were required for the detrital material included in the limestones, the supply of carbonate brought into the Cordilleran Sea from the open oceans, and the more favorable conditions for deposition in the Cordilleran Sea than elsewhere. His final result for the duration of the Paleozoic was:

Cambrian detrital sedimentary rocks	0.5 million years
Post-middle Cambrian sedimentary rocks	0.66 million years
Paleozoic limestone	16.3 million years
Total for Paleozoic time	17.5 million years

Walcott considered the 17.5 million years a minimum value, because he had chosen all of the data so that his result would be conservative.

So far, so good, but the Paleozoic represents only a fraction of geologic time. Various authors had estimated the relative lengths, in arbitrary numbers, of the Cenozoic, Mesozoic, and Paleozoic eras, based on the cumulative thicknesses of sedimentary rocks. Among them were James D. Dana in 1875 (Cenozoic 1, Mesozoic 3, Paleozoic 12) and Henry S. Williams in 1893 (Cenozoic 1, Mesozoic 3, Paleozoic 15). Walcott, however, felt that the estimates of Dana and Williams overestimated the Paleozoic, so he adopted relative lengths of 2, 5, and 12 for the Cenozoic, Mesozoic, and Paleozoic, respectively. For the Precambrian, Walcott estimated that the Proterozoic was about equivalent to the Paleozoic, and that 10 million years was a fair but highly uncertain estimate for the Archean (the Priscoan had not been proposed at that time). Walcott then summed these values to determine the whole of geologic time:

Cenozoic	2.9 million years
Mesozoic	7.24 million years
Paleozoic	17.5 million years
Algonkian	17.5 million years
Archean	10.0(?) million years
Total	55.14 million years

Taking into account the various uncertainties, he concluded that the age of Earth was between 35 and 80 million years.

The techniques for estimating the age of Earth from sedimentation and erosion were as varied as the scientists who devised them. William J. Sollas, for example, used a global approach. Rather than use the sedimentary rocks in one depositional basin, as Walcott had done, Sollas estimated the thicknesses and deposition rates for all post-Archean rocks, then doubled that to account for the Archean, and added a bit to account for the missing stratigraphic record. His result was a final age for Earth of 80 million years.

Warren Upham (1850–1939), a geologist colleague of Clarence King who spent the last 20 years of his life as an archeologist with the Minnesota Historical Society, tried to determine the duration of the Phanerozoic by multiplying the time that had elapsed since the great ice sheets disappeared from North America by ratios for the remainder of the Phanerozoic. The result was 40–50 million years since the beginning of life. Trying another approach, Upham then estimated the length of the Quaternary period as 100,000 years and, based on the changes observed in fossil animals since the beginning of the Cenozoic era, 3 million years for the length of Cenozoic time. Applying this value to Dana's ratios, he obtained 48 million years for the time since the beginning of the Cambrian period. The diversity of life in the Cambrian suggested a long time before for their development, leading Upham to conclude that "the time needed for the deposition of the earth's stratified rocks and the unfolding of its plant and animal life must be about a hundred millions of years" (Upham 1893, p. 218).

Despite the popularity of the method among geologists and the confidence with which most presented their results, the many errors in using sediment accumulation to determine the planet's age were largely recognized at the time. Probably the most serious flaw in the method, ironically, was the one about which they had few qualms—the assumption of uniform rates of erosion and deposition. These factors are now known to vary so much that it is simply impossible to determine an accurate rate for any period of geologic time. Moreover, the Precambrian, for which the record is both highly incomplete and nearly intractable to detailed analysis, constitutes 87% of geologic time.

Sediment accumulation as a method of determining the age of Earth was no better than the other contemporary methods. But in the end it didn't matter, for waiting in the wings was a family of methods that would eventually yield an answer beyond the imagination of most nineteenth-century scientists.

Radioactivity: The Beginning

For centuries, the dream of the medieval alchemists was to turn base metals into gold. They never succeeded, but at about the turn of the twentieth century, a handful of European scientists made the startling discovery that nature performs such transmutations regularly through the phenomenon of radioactivity. It was this discovery that led to the methods now used to determine the ages of rocks, the planets, and the Solar System.

In 1896, the French physicist Antoine-Henri Becquerel (1852–1908) discovered that uranium (U) salts emitted invisible rays similar to X-rays. Two years later, Marie S. Curie (1867–1934) and her husband Pierre (1859–1906) discovered that thorium (Th) also emitted radiation, and they named this new phenomenon *radioactivity.*

In 1902, the British physicist Ernest Rutherford and the British chemist Frederick Soddy published the results of a series of experiments that led them to formulate a general theory predicting the rates of radioactive change. They would each eventually be awarded the Nobel Prize for their work on radioactivity. They began by isolating the radioactive gas emitted by thorium compounds and observing the change in its activity (disintegrations per second) when left to itself. They discovered that after 54.5 seconds the activity was only one-half of its initial value, after 109 seconds only one-quarter of its initial value, after 163.5 seconds only one-eighth of its initial value, and so forth. They had discovered that the decay of the radioactive gas, later identified as an isotope of radon (^{220}Rn), was exponential. With the passage of each half-life, 54.5 seconds in the case of ^{220}Rn, half of the remaining radioactive isotope decays. A similar experiment showed that another radioactive gas, which they isolated from thorium and was later identified as ^{224}Ra (radium), decayed in a similar manner but with a different half-life.

These observations were similar to those made by Becquerel on uranium the previous year, and led Rutherford and Soddy to propose that the atoms of radioactive elements are unstable and decay spontaneously to other elements with the emission of alpha or beta particles. They also proposed that the activity of a substance is directly proportional to the number of atoms present, a proposition from which the formula for radioactive decay can be derived.

Another important suggestion offered by Rutherford and Soddy was that helium might be the product of the decay of radioactive elements. This was confirmed the following year by William Ramsay and Soddy, who showed that helium is formed by the decay of radium. In 1905, Bertram B. Boltwood examined the composition of naturally occurring uranium minerals. Invariably, he noted, they contain lead and helium. Moreover, there was more lead and helium in the older minerals than in the younger. He concluded that lead might be a decay product of uranium. The stage was now set for the first attempts to apply the newly discovered and poorly understood phenomenon of radioactivity to the problem of geologic time.

In 1904, in a presentation at the International Congress of Arts and Sciences that was later published in 1905, Rutherford offered the possibility of using radioactivity as a geologic timekeeper. He stated: "If the rate of production of helium by radium (or other radioactive substances) is known, the age of the mineral can at once be estimated from the observed volume of helium stored in the mineral and the amount of radium present" (Rutherford 1905, p. 33).

Rutherford offered two examples of the proposed radioactive method for calculating ages. The first was a sample of the mineral fergusonite. Using an estimate for the production rate of helium from uranium and its associated radium, Rutherford calculated that the age of the mineral was 497 Ma. (Ma for million years and Ga for billion years are standard scientific shorthand and will be used throughout the rest of this book.) This age, cautioned Rutherford, was a minimum, because some of the helium had probably escaped. A calculation for a second mineral, a uraninite from Glastonbury, Connecticut, also yielded a minimum uranium-helium age of about 500 Ma. Since isotopes were not discovered until 1914, these were "chemical ages" and subject to errors from which modern isotope ages do not suffer.

ATOMS, ISOTOPES, AND OTHER STUFF

It is not practical to discuss the age of Earth or radioactivity without using a few simple terms that describe atomic structure and behavior. If you are familiar with the terms proton, neutron, isotope, nuclide, and beta decay, feel free to skip this box. If not, then please read on—it's neither long nor scary.

Atoms of elements are composed of a tiny and very dense *nucleus* surrounded by shells of orbiting *electrons*, which have one negative charge each. The nucleus consists of one or more *protons* and a variable number of *neutrons*. These subatomic particles are composed of still smaller particles with odd names like mesons and quarks, but we don't need to concern ourselves with them. Protons and neutrons are each approximately 1836 times as heavy as an electron.

Nearly all of the mass of an atom resides in the nucleus. Since energy equals mass times the speed of light squared ($E = mc^2$), most of the available energy of an atom is in the nucleus. A nucleus occupies only about a trillionth (10^{-12}) of the volume of an atom and has a density of more than 100 trillion times that of water. Protons carry one positive charge each and exactly balance the negative charge of one electron, whereas neutrons carry no charge. Thus, a neutral atom contains an equal number of protons and electrons and has no electrical charge. The addition or subtraction of one or more electrons results in an atom with an electrical charge. This charged atom is called an *ion*. Because of their electrical charge, ions can combine with ions of other elements to form molecules and chemical compounds.

The atoms of any given element always contain a unique number of protons but may contain a variable number of neutrons. For example, hydrogen always contains one proton, but it may contain zero, one, or two neutrons. Each of these different configuations of the hydrogen nucleus is called an *isotope* of hydrogen. All elements have isotopes, although some may no longer occur naturally because they have decayed away. All isotopes of all elements are collectively called *nuclides*, so any isotope of any element is also a nuclide. The sum of the number of protons and the number of neutrons is the *mass number*, which is indicated by a superscript that precedes the element symbol. Thus, hydrogen with two neutrons is ^{3}H (also called tritium), potassium (19 protons) with 21 neutrons is ^{40}K, uranium (92 protons) with 143 neutrons is ^{235}U, and so forth.

Too many neutrons in the nucleus causes a nuclide to be radioactive, which is a way for an atom to shed excess energy and become more stable. How many neutrons is too many depends on the particular element. For instance, ^{3}H, ^{40}K, ^{235}U are all radioactive. Most radioactive decay involves a change in the number of protons, so the radioactive nuclide decays into an isotope of a completely different element. For example, ^{40}K decays to ^{40}Ar (argon), ^{87}Rb (rubidium) decays to ^{87}Sr (strontium), and so on.

Each radioactive nuclide decays in a particular way, and only three types of radioactive decay are important in the radiometric (or isotope) dating of rocks. In *beta decay*, a neutron breaks up into a proton and an electron, and the electron, known as a *beta particle*, is ejected from the nucleus. In *alpha decay*, the nucleus fragments and an *alpha particle*, which is a helium nucleus (^{4}He) and contains two neutrons and two protons, is ejected from the nucleus. In *electron capture decay*, an electron from the innermost electron shell falls into the nucleus, where it combines with a proton to create a neutron. Each radioactive nuclide decays at a unique rate called its *half-life*, which is the time it takes for exactly half of the remaining radioactive atoms to decay. Half-lives range from a fraction of a second to billions of years.

There. That's all the nuclear physics you need to read this book.

The same year that Rutherford's results appeared in print, Robert J. Strutt, the son of Lord Raleigh, who won the Nobel Prize in Physics in 1904, measured the uranium, thorium, radium, and helium content of twenty-two radioactive minerals. Strutt found that when helium was abundant, the mineral also contained thorium, but the converse did not hold. This strengthened the conclusions that thorium, in addition to uranium and radium, produced helium, and that the helium quantity was a function of age.

In 1908, Strutt compared the ratio of helium to uranium in thirteen samples of phosphate nodules and phosphatized bone with their geologic ages. He found that the ratios of helium to uranium did not uniformly follow the known relative ages, but also that high ratios did not occur in the younger samples. He hypothesized that helium was imperfectly retained, but he calculated minimum ages ranging from 0.225 to 141 Ma for four of the samples.

In 1905, Rutherford was invited to deliver the Silliman Lectures at Yale University. In his talks, Rutherford suggested that age calculations based on lead might be superior to those based on helium, because the lead could not escape. This idea was first put to the test by Boltwood in 1907, who compiled analyses of forty-three uranium-bearing minerals from ten different localities. Using the average ratio of lead to uranium for each locality, an estimate of the decay constant of uranium, and a formula suggested by Rutherford, Boltwood calculated the ages for each locality. Boltwood was cautious about the significance of his ages, but concluded that they might be of value for determining the ages of some types of geologic formations.

In 1911, Arthur Holmes, a young British geologist who had just taken a position with Memba Minerals prospecting in Mozambique, reevaluated Boltwood's data, added new analyses of uranium-bearing minerals from Norway, and calculated uranium-lead ages for nine localities using an improved estimate of the uranium decay constant (Table 3.2). Holmes concluded:

> Wherever the geological evidence is clear, it is in agreement with that derived from lead as an index of age. Where it is obscure, as, for example, in connection with the pre-Cambrian rocks, to correlate which is an almost hopeless task, the evidence does not, at least, contradict the ages put forward. (Holmes 1911, pp. 255–256)

These first radiometric ages by Rutherford, Boltwood, Strutt, and Holmes were chemical ages. They were calculated before isotopes were discovered, before the decay rates and intermediate decay products of uranium were known, and before it was discovered that lead is also produced by the decay of thorium. These factors combined to produce ages that were usually too high. For example, modern uranium-lead isotope ages of samples from the Glastonbury, Branchville, and Spruce Pines localities show that Boltwood's ages were excessive by 120–170 million years, and Holmes's by 20–65 million years (see Table 3.2). Ages based on helium were further complicated by helium loss, a problem that renders uranium-helium methods largely useless to this day.

For all their imperfections, these early and highly experimental ages based on the decay of uranium were at least as firmly grounded in both theory and empirical evidence as those methods that relied on the cooling of Earth, the accumulation of sodium in the oceans, or the rates of sedimentation and erosion. Radioactivity, at least, proceeds at a rate that is constant over time

Table 3.2
A Comparison of Boltwood's and Holmes' Chemical Uranium-Lead Ages with Geologic Ages Determined by Holmes

Locality	Geologic Age	U-Pb Age (million years)	
		Boltwood (1907)	Holmes (1911)
Glastonbury CT	Carboniferous	410	340
Norway	Devonian	—	370
Spruce Pine NC	pre-Carboniferous	510	410
Marietta SC	pre-Carboniferous	460	410
Branchville CT	Silurian-Devonian	535	430
Sweden and Norway	Precambrian	1300	1025
Sweden and Norway	Precambrian	1700	1270
Texas	Precambrian	1800	1310
Colorado	Precambrian	1900	1435
Ceylon	Precambrian	2200	1640

NOTE: The difference between the two sets of ages is primarily due to the use of different decay constants.

SOURCE: Data from Boltwood 1907 and Holmes 1911.

and under all conditions (well, almost all, but more on that later). Although these first radiometric mineral ages did not directly date the time of Earth's origin, their importance to scientific thought about the age of the planet cannot be overestimated. They were the first quantitative indication, based on physical principles rather than scientific intuition, that Earth might be billions, rather than a few tens or hundred millions, of years old. The discovery of radioactivity and the heat it produced had made Kelvin's simple conductive cooling calculations obsolete. Now that same phenomenon was providing quantitative evidence that those geologists who had insisted that Earth must be very, very old might be right.

In addition to providing methods for determining the ages of rocks, radioactivity also offered the possibility of calculating the age of Earth from the relative abundances of the radioactive elements and their decay products. The first such calculation appeared in 1921 in a paper entitled "A Superior Limit to the Age of the Earth's Crust" by Henry N. Russell, Professor of Astronomy at Princeton University. Using estimates of the amount of uranium and lead in Earth's crust, Russell calculated that the lead would be formed from the decay of uranium and thorium in 8 billion years. This was an upper limit, however, because lead may have been present initially in the crust. Moreover, speculation existed that uranium itself might be produced in the crust by the decay of some other element. The lower limit for the crust, Russell noted, must be considerably greater than 1.1 billion years, which was the approximate age of the oldest Precambrian minerals that had been dated by uranium-lead. He concluded:

> Taking the mean of this and the upper limit found above from the ratio of uranium to lead, we obtain 4×10^9 years as a rough approximation to the age of the Earth's crust. . . . Indeed, it might be safe to say that the age of the crust is probably between two and eight thousand millions of years. (Russell 1921, p. 86)

In a popular 1927 booklet entitled *The Age of the Earth: An Introduction to Geological Ideas*, Holmes revised Russell's calculation using current estimates of crustal composition. He found that the age of Earth from this calculation

> is just over 3,000 million years. The age of the earth cannot exceed this figure, because some of the lead in rocks may be ordinary lead of atomic weight 207.2, and because the radio-active elements may have existed in the sun and have there generated lead before the earth was born. We should therefore expect the age of the earth to be nearer 1,600 million years than 3,000 million years. (Holmes 1927, p. 72)

The lower limit of 1600 Ma Holmes obtained by adding the approximate length of a geologic time scale era (300–400 million years) to the age of the oldest dated mineral (then 1260 Ma, from western Australia).

At the time Holmes published his 1927 booklet, there were age data available for only twenty-three localities and Holmes was able to summarize them all in a brief table. Yet the ages, ranging from 35 to 1260 Ma, were so consistent with the geologic ages of the localities that they were difficult to doubt. In addition, if rocks in Earth's crust were more than 1 billion years old, then Holmes's age range of 1.6–3.0 Ga for Earth was credible. In his final chapter, Holmes tabulated the physical evidence for the age of Earth and concluded that it was consistent with the age of 1.6–3.0 Ga based on radioactivity. Results from sodium accumulation and sediment thickness were relegated to insignificance; cooling calculations were not even listed. The methods so important to Kelvin and the other pioneers in the search for Earth's age had been rendered obsolete by the new evidence from the decay of radioactive elements—the short geologic time scale was disproved once and for all.

The discovery of new radioactive elements and the measurement of their rates of decay in the period between 1930 and 1950 provided the means for additional estimates of the age of Earth or, more exactly, of Earth's matter. The methods all followed more or less the general approach established by Russell and involved estimating the crustal or earthly abundances of a radioactive parent nuclide and its ultimate decay product, then calculating the time required for all of the product nuclide to be generated by decay of the parent. It was, however, a method based on Russell's original idea—the decay of uranium to lead—that finally provided the best value for the age of Earth and the Solar System. That story will be told in Chapter 8.

CHAPTER FOUR

Clocks in Rocks: How Radiometric Dating Works

When you want to determine time or dates in the everyday world, you consult a watch, a calendar, or a written record of events. But rocks and minerals have no clocks or time record, or do they? They do—built-in atomic clocks in the form of the long-lived radioactive isotopes of elements that nearly all rocks and minerals contain. These atomic clocks are based on the spontaneous, inexorable, and constant decay of radioactive isotopes of elements, called *parent isotopes,* into stable (nonradioactive) isotopes of other elements, called *daughter isotopes.* Each parent-daughter pair constitutes an independent clock that provides the information necessary to calculate the time that has elapsed since the rock or mineral formed. The use of these parent-daughter pairs of isotopes to tell geologic time is called *radiometric dating,* or isotope dating.

A number of long-lived radioactive isotopes are used in radiometric dating, and there are several ways in which they are used (Table 4.1). Each technique is usually referred to by the chemical symbols of the parent-daughter pair. Thus, K-Ar is the potassium-argon method, U-Pb is the uranium-lead method, and so forth. There are a few variations on this convenient form of notation that will become apparent later. Each radiometric method has unique characteristics that make it applicable to particular rocks, particular minerals, or particular geologic circumstances. This means that not all the methods can be used on all rocks or under all conditions. It also means that the methods are complementary rather than redundant, with each method working best under somewhat different geologic circumstances. All radiometric methods, however, have two things in common: (1) they utilize isotopes of elements that occur in measurable quantities in common rocks and minerals, and (2) the half-lives of the parent nuclides are sufficiently long that they can measure time in millions and billions of years.

Table 4.1
Radiometric Dating Methods used for Determining the Ages of Rocks and Minerals

Method	Parent Isotope (radioactive)	Daughter Isotope (stable)	Half-Life (billion years)
K-Ar (potassium-argon)	^{40}K	^{40}Ar	1.25
Ar-Ar (argon-argon or $^{40}Ar/^{39}Ar$)	^{40}K	^{40}Ar	1.25
Rb-Sr (rubidium-strontium)	^{87}Rb	^{87}Sr	48.8
Sm-Nd (samarium-neodymium)	^{147}Sm	^{143}Nd	106.
Lu-Hf (lutetium-hafnium)	^{176}Lu	^{176}Hf	35.9
Re-Os (rhenium-osmium)	^{187}Re	^{187}Os	43.0
Th-Pb (thorium-lead)	^{232}Th	^{208}Pb	14.0
U-Pb (uranium-lead)	^{235}U	^{207}Pb	0.704
U-Pb (uranium-lead)	^{238}U	^{206}Pb	4.47
Pb-Pb (lead-lead)	^{235}U and ^{238}U	^{207}Pb and ^{206}Pb	0.704 and 4.47

NOTE: Superscripts indicate the mass number (neutrons + protons) of the isotopes.

Radioactivity as a Timekeeper

Even though the physical mechanisms governing radioactivity are complex, the general concept of how decay proceeds is relatively simple. Radioactive decay is a statistical process in which each atom of a given radioactive nuclide has exactly the same probability of decaying in some particular period of time as any other atom of that same nuclide. That characteristic probability is known as the *decay constant,* which is the probability that a decay will take place in a specified time (second, year, etc.). Decay constants range from a low of zero for a stable (nonradioactive) nuclide to a high of 1 for a nuclide that decays the instant it is formed. For example, suppose that in a jar there are 100 atoms of a radioactive nuclide with a decay constant of 0.1 per year. This means that on average, we can expect 10 of the atoms (10%) to have decayed by the end of the first year, 9 (10% of the remainder of 90) at the end of the second year, 8 (10% of the remainder of 81) at the end of the third year, and so on. Each decay results in the creation of one daughter atom from one parent atom, so there is no net change in the total number of atoms.

In a way, the use of parent and daughter nuclides to date rocks is analogous to measuring time using an hourglass. Sand grains in the top chamber of the hourglass run through a small hole into the bottom chamber. The size of the sand grains, the size of the hole, and gravity precisely control the rate of flow. In an hourglass, however, there is no change in the type of sand grains, only in their location. Still, there is a known rate of change, and elapsed time could be figured by counting the sand grains in either the top or the bottom chamber, even though it is usual to just let the sand run out. There is another difference as well. The number of sand grains in the hourglass that change location is linear, that is, the same for any interval of time, whereas the number of radioactive decays that occur in any given time follows an exponential curve. Let's see why that is.

When it comes to decay, radioactive nuclides are independent creatures and pay no attention to what other nuclides of their ilk are doing. Because of this independence, the more radioactive atoms there are the more decays occur. It's not that they influence each other, it's just that there are more of them. Thus, the number of radioactive atoms that decay in any given period of time is directly proportional to the number of radioactive atoms present. As the number of decaying atoms decreases, the number of decays in a given period of time slows proportionately. There is a corresponding decrease in the rate of growth of the number of daughter atoms as well. This type of change, where the rate depends on the number present, is called *exponential change.* Radioactivity is not the only phenomenon where the rate of growth (or decrease, which is simply negative growth) is directly related to the size of the growing quantity. The growth of the world's population is also exponential—the more people there are, the faster the population grows, provided that the birth rate remains constant.

Because decay is a statistical process, it is not possible to tell exactly when any particular atom will decay. For a small number of atoms, therefore, it is virtually impossible to determine the exact number of decays in a given time. In the case of the 100 atoms in a jar discussed above, 13 might decay the first year, 6 the second, 8 the third, and so on. For large numbers of atoms, however, the statistical uncertainty is negligible and the number of atoms that will decay in a given period of time is precisely predictable. Fortunately, the numbers of atoms in even small amounts of matter are very large—as little as one one-hundred-thousandth (0.00001) of a gram of potassium contains 150,000 trillion (1.5×10^{17}) atoms—so the statistical nature of radioactive decay is of no practical concern to the accuracy of radiometric dating.

The term *half-life,* which was introduced in Chapter 3, is related to the decay constant in a simple way—it is just the logarithm of 2, which is 0.693, divided by the decay constant. For the hypothetical isotope represented by

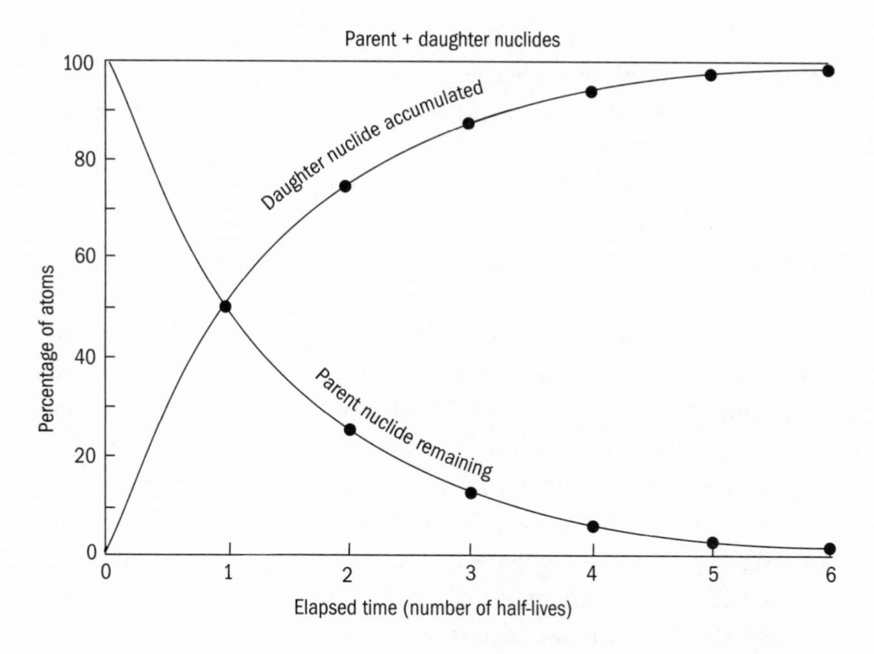

Figure 4.1 The decay of a radioactive parent nuclide and the corresponding accumulation of its daughter nuclide. In a closed system where neither parent nor daughter escapes, the sum of the parent and daughter nuclides at any time equals the original amount of the parent nuclide.

the 100 atoms with a decay constant of 0.1 per year, the half-life is 6.93 years. This means that every 6.93 years, one-half of the remaining atoms will decay and a corresponding number of daughter atoms will be created.

When the percentages of the parent and daughter nuclides are graphed as a function of the number of elapsed half-lives, the result is two exponential curves—one showing the decrease of the parent and the other the increase of the daughter (Figure 4.1). If none of the parent or daughter atoms escapes, then the sum of the two curves (parent + daughter) must always equal the original number of parent atoms; thus, their sum plots as a straight line in the graph. After one half-life, 50% of the atoms will be parent atoms and 50% will be daughter atoms, after two half-lives the ratio of parent to daughter will be 25% to 75%, and so on, with the total always being 100%. This predictable relationship between parent atoms and daughter atoms in a closed system is the foundation of radiometric dating. For radiometric dating, the system, usually a rock or specific mineral grains, need only be closed to the parent and daughter nuclides. Because the number of parent and daughter atoms must always equal the original number of parent atoms present, the age equation can be expressed entirely in terms of quantities that can be measured in the rock today—the parent, the daughter, and the decay constant. This leaves time as the only unknown, a quantity for which the equation is easily solved.

If the rock or mineral to be dated contained none of the daughter nuclide at the time of formation, at any later time, the age of the rock or mineral could be found from the measurements and the simple relationship described above. If the rock or mineral incorporated some of the daughter nuclide when it formed, however, then this initial amount of the daughter would have to be subtracted from the total amount of daughter measured in order to calculate the correct age. It may seem that the requirement of knowing the amount of initial daughter present in a rock or mineral is a formidable limitation to the accuracy of radiometric dating. As we shall see, however, for the principal methods the amount of the initial daughter is either zero, negligible, or not required in the age calculations.

Is Decay Constant?

One of the primary requirements of a radiometric clock is that decay is constant and predictable since the clock was first "set," that is, from the time the rock or mineral was formed to the time the measurements are made. Just how reasonable is this hypothesis? The answer is that unless there has been some undiscovered change in the fundamental nature of matter and energy since the Universe formed, the presumption of constancy for radioactive decay is, for all practical purposes, eminently reasonable.

There are two reasons that significant changes in the rates of radioactive decay do not occur. The first is that the nucleus of an atom is extremely small and well insulated by its cloud of orbiting electrons. These electrons not only separate atomic nuclei by relatively great distances so that the nuclei cannot interact, but they also provide a "shield" that prevents ordinary chemical or physical factors from affecting the nucleus. Chemical activity in an atom, for example, occurs almost entirely in the outermost electrons and does not involve the nucleus at all.

The second reason is that the energies involved in nuclear changes are a million times greater than those involved in chemical activity and 10,000 to 100,000 times greater than the energies that bind the electrons to the nucleus. This is the reason that nuclear reactors and powerful particle accelerators are required to penetrate and make changes in atomic nuclei. Except in nuclear reactions, such energies are generally unavailable in natural processes, including those that form, change, and affect rocks on Earth and in the Solar System.

Not long after the discovery of radioactivity, experiments were conducted to determine whether or not the rates of radioactive decay could be changed. In 1907, Lord Rutherford and his colleague, J. E. Petavel, placed a sample of what was then known as "radium emanation" (^{220}Rn) in a steel-encased bomb. They observed no change in the activity of the sample even though the explosion involved an estimated temperature of 2500° C and pressure of 1000 atmospheres. Madame Curie and M. Kamerlingh Onnes found in 1913 that lowering the temperature of a radium compound to the boiling point of liquid hydrogen (–252.8° C) did not change the radium activity more than 0.05%, if at all. Other experiments involved varying gravity by changing locations from mountain tops to the depths of mines or by whirling samples in a centrifuge, and applying strong magnetic fields. These early experiments and subsequent ones, involving extremes of temperature, pressure, chemical state, and electrical and magnetic fields, have uniformly failed to induce any detectable changes in the decay rates of a wide variety of radioactive nuclides that decay by alpha or beta decay. These observations are in accord with modern theory, which predicts that changes in alpha and beta decay rates, although possible, should be much less than 0.01%.

In contrast, small changes in the rates of electron-capture decay have been observed. This is not surprising, because this type of decay involves an electron from the innermost electron shell. If the distance of the electrons from the nucleus could be changed, then a change in the probability that an electron would be captured by the nucleus might be expected, and this does happen.

The possibility of such an effect was predicted in 1947 by the physicist and Nobel laureate Emilio Segré, who suggested that changes in the decay rate of ^{7}Be (beryllium) might be most easily induced because it has very few electrons and they are quite close to the nucleus. The activity of ^{7}Be has been decreased by combining beryllium with other elements into different chemical compounds. For example, there is a 1.7% difference in the half-life of ^{7}Be in the pure element and in beryllium oxide.

An effect of pressure on the decay rate of ^{7}Be in beryllium oxide has also been reported, but it amounts to an increase of only 0.6% at a pressure equal to that at a depth of about 750 km below Earth's surface.

Another circumstance that can change decay rates is if all of the electrons are removed from an atom, leaving only the bare nucleus. When ^{187}Re (rhenium) is stripped of its 75 electrons, its half-life plummets from 42 billion years to only 33 years. Dysprosium, or ^{163}Dy, is normally a stable nuclide, but when its 66 electrons are removed it becomes radioactive with a half-life of 47 days. Removing all the electrons from an atom, however, is a drastic act. It can be done in the laboratory, but in nature it occurs only in the hot interiors of stars where the temperatures are extreme. Therefore, this mechanism may be important in the creation of elements in stars and supernovae, but it is not relevant in radiometric dating.

In summary, both theory and experiments show that changes in decay rates are not only rare, but, when they do occur, small. Even the largest observed change of 1.7% in ^{7}Be would have a relatively small effect on a measured radiometric age, but ^{7}Be is not used for dating. Although changes in the decay rates of a few elements have been observed under special circumstances, no changes have ever been detected in any of the nuclides used for dating, and none of significance is theoretically expected. Thus, we can be confident that the radiometric clocks used for geologic dating "tick" at rates that are, for all practical purposes, unchanging.

Measuring Isotopes

We'll detour here, just briefly, to consider how isotopes are measured with such precision that scientists can determine ages of rocks to within a few percent or better. In 1914, James J. Thomson of the Cavendish Laboratories at Cambridge University invented an instrument called the parabola mass analyzer. Thomson was a friend of Lord Kelvin (though no relation) and an accomplished and well-known scientist of his day. Among other things he discovered the electron, for which he was awarded the Nobel Prize in Physics in 1906. He also found the first two isotopes (of neon) using his new mass analyzer, and observed that either potassium or sodium—he couldn't tell which, but it was later found to be potassium—was radioactive. Thomson's parabola mass analyzer was the forerunner of today's *mass spectrometer*, which is used to measure isotopes precisely; it comes in a variety of sizes, from tiny ones that fly on spacecraft to large ones that fill a room.

The principle of a mass spectrometer is rather simple. Atoms are introduced at one end, where they are stripped of one or more of their electrons by a focused beam of particles, usually other electrons, which turns the atoms into ions. Because they have an electrical charge, the ions can be manipulated by electrical and magnetic fields. The ions are focused and accelerated down a tube toward the other end using electrical fields. As they travel down the tube, the ions are bent by a strong magnetic field, the lighter ones being bent sharply and the heavier ones being bent less, thus separating the ions according to their mass. By changing the strength of the magnetic field, each mass beam can be directed into a detector, which precisely measures the quantity of ions present relative to the other mass beams. All of this happens in a very high vacuum so the ions don't collide with stray atoms not of immediate interest.

Whereas Thomson's parabola mass analyzer and its immediate successors used photographic plates as detectors, today's mass spectrometers use electronic detectors of various sorts, some of which can amplify the ion beam signals by many orders of magnitude and others that can literally count the number of ions present. In addition, modern instruments are invariably run by computers that not only optimize the conditions for the measurements but also collect the data and do the laborious calculations in an instant. One such instrument, a SHRIMP ion probe, is shown in Figure 4.2. (SHRIMP stands for sensitive high-resolution ion microprobe; it was developed at the Australian National University in Canberra.) It is a specialized type of mass spectrometer that uses a beam of ions, usually oxygen, to blast other ions off of a small spot on the surface of a crystal. The SHRIMP is capable of measuring U-Pb and Pb-Pb ages from very tiny spots (about 0.02 mm) within a zircon crystal. Ar-Ar dating is often done using a laser to heat or melt a small mineral grain (0.5 mm or less), releasing the argon for analysis of its isotopes using a high-sensitivity mass spectrometer designed specifically to measure inert gases. Other types of dating, such as Rb-Sr, Sm-Nd, and so forth, are done using mass spectrometers that analyze small concentrates of the relevant elements after separation in the laboratory by chemical meth-

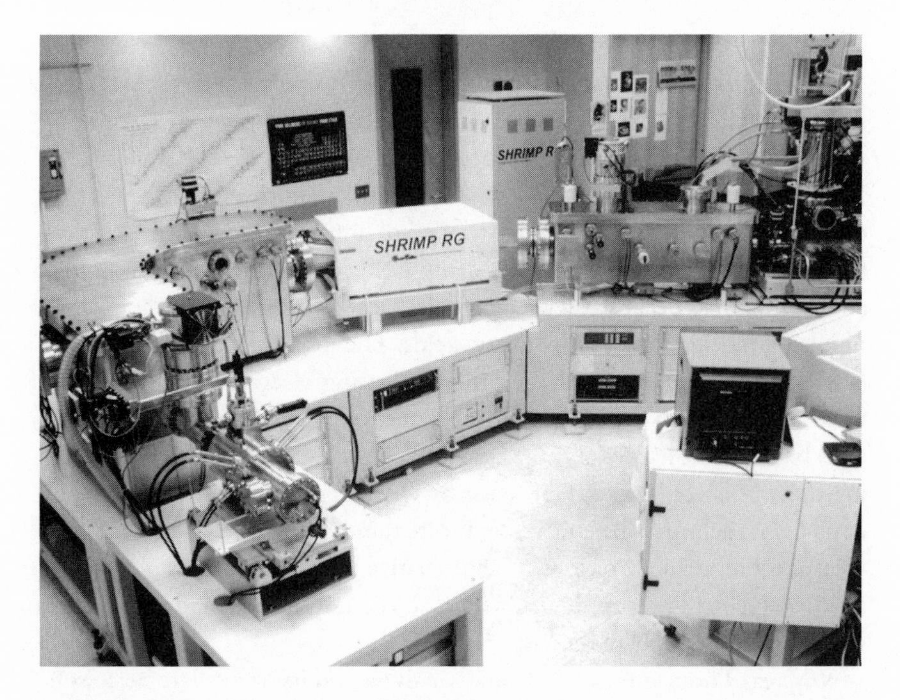

Figure 4.2 The SHRIMP ion microprobe operated by Stanford University and the U.S. Geological Survey. A mineral grain is placed in the far end of the instrument, where a tiny spot on its surface is bombarded by energetic oxygen ions. The oxygen ions blast ions of other elements off of the surface of a crystal. The ions are accelerated down the vacuum tube, separated according to mass by magnetic and electrical fields, and measured by the detectors in the foreground. Such an instrument costs millions of dollars. It is ironic that often the smallest things require the largest instruments to measure them. (Photo courtesy of Brad Ito, U.S. Geological Survey.)

ods. Today's modern mass spectrometers are so precise and so reliable that making accurate isotope measurements, exceedingly difficult only a few decades ago, is now more or less routine.

Simple Accumulation Clocks

In principle, the age of a rock or mineral could be found by measuring the number of parent and daughter isotopes present in a single sample for any of the dating methods listed in Table 4.1. In practice, however, the nearly ubiquitous presence of an unknown amount of initial daughter isotope generally prevents the use of such simple accumulation methods. The exceptions are the K-Ar method and special but rare cases of the other methods. This "initial daughter" problem is easily solved by the application of some clever graphical methods, which are discussed later in this chapter.

The K-Ar method is the only decay scheme that can be used with little concern for the initial presence of the daughter isotope. This is because ^{40}Ar is an inert gas that does not combine chemically with any other element and can escape from rocks when they are heated. Thus, while a rock is molten, the ^{40}Ar formed by the decay of ^{40}K escapes from the rock liquid. After the rock has solidified and cooled, the ^{40}Ar is trapped within the crystal structures of the minerals like a bird in a cage and accumulates there with the passage of time. If the rock is heated or melted at some later time, then some or all of the ^{40}Ar may escape, and the K-Ar clock is partially or totally reset. Some cases of initial ^{40}Ar have been found, but they are uncommon and the amounts usually so small that they would have a negligible effect on all but the youngest of K-Ar ages.

Like all radiometric methods, the K-Ar method does not work on all rocks and minerals under all geologic conditions. The K-Ar clock works particularly well (and is widely used) on igneous rocks, such as lava flows, that form from molten rock and have not been heated significantly since their formation. It, along with all radiometric clocks, does not work on most sedimentary rocks because these rocks are composed of debris from older rocks. It also does not work well on many metamorphic rocks because this type of rock forms from other rocks under heat and pressure but without undergoing complete melting. Many metamorphic rocks have complex histories involving several heatings, none of which may be sufficient to release all of the accumulated argon. As a result, it is difficult to know when or if the K-Ar clock in a metamorphic rock was last completely reset.

The Rb-Sr method is based on the radioactivity of ^{87}Rb, which undergoes beta decay to ^{87}Sr. Rubidium never forms minerals of its own, but its chemistry is similar to that of potassium and sodium, both of which do form many common minerals, so rubidium occurs as a *trace element* in most rocks where it substitutes for potassium and sodium. The chemistry of strontium, also a trace element, is similar to that of calcium, a major element in rocks, for which strontium may substitute.

Unlike argon, which escapes easily and entirely from most molten rocks, strontium is present as a trace element in most minerals when they form.

For this reason, simple Rb-Sr accumulation ages can be calculated only for those rare minerals that are high in rubidium and contain a negligible amount of initial strontium. The Rb-Sr method is very useful on rocks with complex histories because strontium is chemically bound within the crystals and does not escape from minerals nearly as easily as argon. As a result, a sample can obey the closed-system requirements for Rb-Sr dating over a wider range of geologic conditions than a sample for K-Ar dating.

Within the past decade or so, improvements in the precision and sensitivity of analytical techniques have enabled scientists to exploit three decay schemes that were previously of little value. These include the decay of ^{147}Sm to ^{143}Nd, ^{176}Lu to ^{176}Hf, and ^{187}Re to ^{187}Os. The combination of the long parental half-lives and the very low natural abundance of both parent and daughter isotopes means that the quantities of the parent and daughter isotopes in most rocks and minerals and the changes in the amounts of accumulated daughters over time are both very small. Nevertheless, these three decay schemes have some unique advantages not found in the more commonly used methods.

Samarium and neodymium are chemically similar, and ^{147}Sm decays by alpha decay to ^{143}Nd with a very long half-life (see Table 4.1). These elements occur in trace amounts in nearly all rocks and minerals although their concentrations are usually only a few parts per million or less. Because of their chemical similarity, natural geochemical processes do not cause extreme fractionation of samarium and neodymium, and the Sm-Nd method is more resistant to disturbances caused by metamorphism than other dating methods. This is a decided advantage for age measurements of very old rocks whose histories are complex.

The Lu-Hf method is based on the beta decay of ^{176}Lu to ^{176}Hf. Lutetium is chemically similar to samarium and neodymium, but hafnium is not. Both occur as trace elements in most rocks but usually in concentrations of less than 1 part per million (ppm).

The Re-Os method is based on the beta decay of ^{187}Re to ^{187}Os. Both rhenium and osmium are metals whose average abundance in igneous rocks is very small. Because of the extremely low abundance of rhenium, the method is generally not applicable to common rocks and minerals. It has proven of value, however, in dating the metal found in some meteorites, where the concentration of rhenium may be as much as 0.5 ppm, whereas that of the other parent isotopes in Table 4.1 is negligible.

The U-Th-Pb methods are based on the radioactivity of ^{235}U, ^{238}U, and ^{232}Th, all of which decay to different isotopes of lead. These three decays differ from the others in Table 4.1 because they each involve a decay series with several intermediate radioactive daughter products, all with short half-lives. The decay series of ^{238}U, for example, includes thirteen intermediate radioactive daughter nuclides of eight elements between ^{238}U and the final stable daughter isotope, ^{206}Pb. The ^{238}U decays to ^{234}Th, which decays to ^{234}Pa (proactinium), which decays to ^{234}U, and so forth, until ^{206}Pb is finally produced by the decay of ^{210}Po (polonium).

Despite the existence of intermediate radioactive daughter isotopes, the decays of uranium and thorium to lead can be treated as if they were a simple, one-step decay. This is because of two fortunate circumstances. The first is that none of the intermediate daughter isotopes occurs in more than a single decay series. Thus, each of the three series ultimately results in a unique isotope of lead. The second is that the half-lives of the intermediate daughters are very much shorter than those of the three parents, so that equilibrium is established very shortly after a new rock is formed. Once the series is in equilibrium, the production rate of lead is exactly equal to the decay rate of the uranium or thorium. The length of time necessary for the various intermediate daughters to reach equilibrium within the decay series can actually be used as a dating tool for young rocks (a few hundred thousand years old or less), but these so-called disequilibrium dating methods will not be discussed because they are not relevant to the age of Earth.

In Chapter 3, some early attempts to use the decay of uranium and thorium to the inert gas helium as a dating method were discussed. The production of helium cannot be reliably used for dating because a helium atom is very tiny and, unlike the larger argon atom, escapes easily from rocks and minerals. The escape of helium, however, has no effect on the accumulation of lead because the helium is a by-product of the decays—it is simply the alpha particles that have acquired electrons—and not a radioactive member of the decay series.

A few common minerals contain significant amounts of uranium and thorium. These minerals, of which zircon is the most common, do not occur in large volume but do occur in many rocks. Zircon is also exceedingly low in initial lead, so the U-Pb and Th-Pb methods can be applied to this mineral and a few others with little concern for initial lead.

Three independent age calculations can be made from the U-Pb and Th-

Pb decays. If these age calculations agree, then that age represents the age of the mineral. More often than not, however, the three ages do not agree. This is primarily because lead is a volatile element that can be lost if the mineral is reheated at some later date. Calculating an age based on the ratio of any pair of the three lead isotopes can minimize the problem of lead loss because the lead isotopes will be lost in proportional amounts. Most often, the ratio of ^{207}Pb to ^{206}Pb is used so that the differences in the chemical behavior of uranium and thorium are completely eliminated. In practice, the Th-Pb method is rarely used because the U-Pb and Pb-Pb methods are superior.

Age Diagnostic Diagrams

We have seen that most of the simple accumulation methods are encumbered with two requirements: (1) that the amount of initial daughter be either negligible or known, and (2) that the rock remains a closed system since formation. But is it possible to find the quantity of initial daughter isotope or know if the system has been open or closed? Yes. These seemingly formidable problems can be solved by the use of simple graphs that are collectively called *age diagnostic diagrams.* These diagrams not only provide an age, but some provide an exact measure of the initial daughter, some provide an age for systems that have not remained closed, and all are self-checking, meaning that they inherently contain information about the reliability of the results. They are especially useful on old rocks with complex histories. In practice the age is found mathematically, but the diagrams are usually used to show the quality of the data and the results of the calculations. These diagrams are important and worth the time needed to understand how they work. But don't worry—there will be no math!

THE IMPORTANCE OF RATIOS

The use of ratios to reveal important relationships is common throughout science. Radiometric dating relies on the measurement of isotope ratios. To understand why, consider an example that is somewhat simpler than isotopes in rocks— variations in the female populations of selected European countries. If the number of females in these countries is graphed, there is a large scatter of data (Figure 4.3a). Technically, there is nothing wrong with this graph; it is an entirely accurate representation of the 1998 census data.

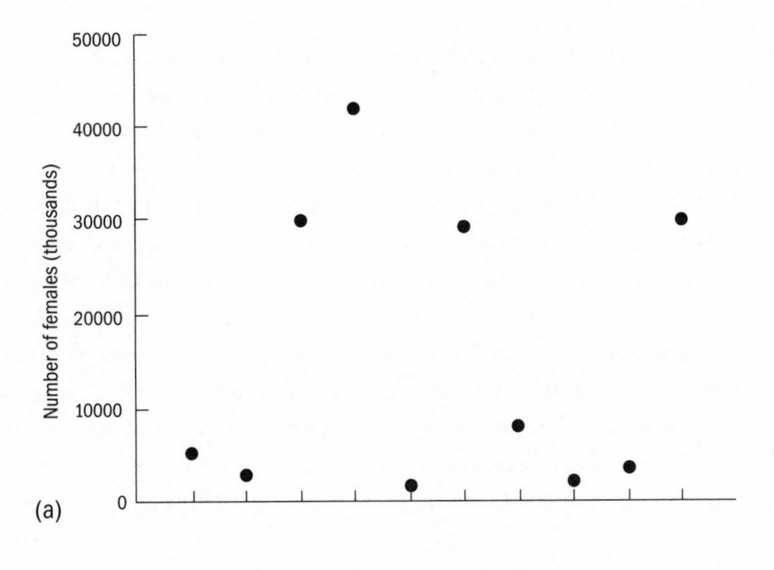

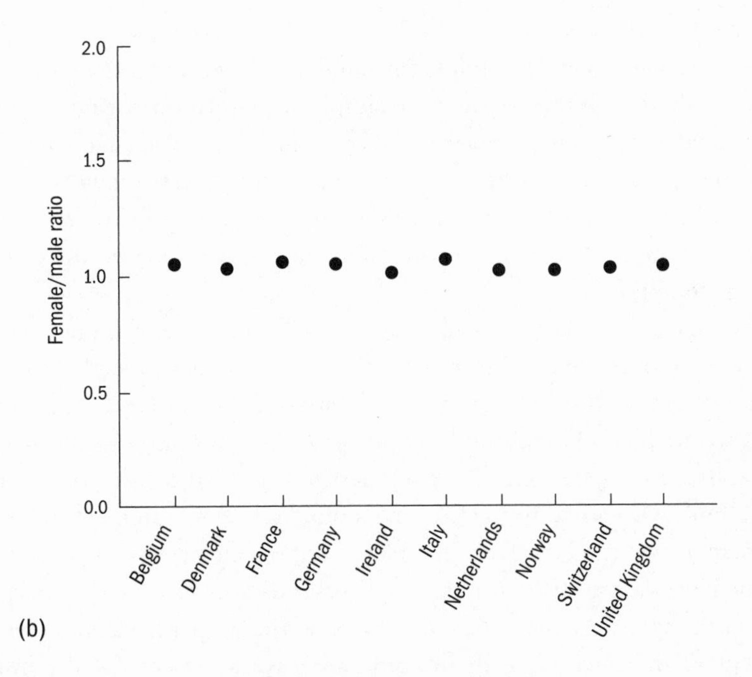

Figure 4.3 Variations in the female populations of selected European countries in 1998. (a) Number of females. (b) Ratio of females to males. Both graphs are valid, but they each reveal different things about the populations of the countries. Ratios (graph b) are often more instructive than absolute numbers (graph a). (After York and Dalrymple 2000.)

But the large variation among countries might have several causes. It could be a function of the populations of the countries or of differences in the birth or death rates of women. How could that hypothesis be tested? For that purpose, Figure 4.3a is not very useful.

If the number of females is divided by the number of males, however, a clear straight-line relationship is revealed (Figure 4.3b). This graph makes it obvious that the ratio of females to males in the populations of these countries is slightly greater than 1, and that variation from country to country is small. The ratios in Figure 4.3b are much more instructive than the numbers in Figure 4.3a. This is only one example where ratios are more useful than absolute amounts. Isotope ratios, rather than isotope amounts, are used exclusively in age calculations and in diagrams that portray radiometric ages because the ratios contain the relevant information about the ages of rocks.

SIMPLE ISOCHRONS

The isochron diagram is a device of magnificent power and simplicity. It has two significant advantages over the simple accumulation dating method. First, it circumvents the problem of the amount of the initial daughter; that information need not be known—it is one of the answers provided by the method. Second, the method is self-checking, providing the user with information about the degree to which the samples have obeyed the closed-system conditions.

As the name implies, an *isochron* is a line of equal time. It is obtained by analyzing several minerals from the same rock, or several rocks that formed at the same time from the same parent material. The isotope data from the analyses are plotted on a simple graph, with the parent isotope on the abscissa (horizontal axis) and the daughter isotope on the ordinate (vertical axis), both normalized to (divided by) a nonradiogenic isotope of the same element as the daughter. If the samples have been closed systems since they formed, the data will fall on a line—an isochron—whose slope is a direct result of the age of the rock; the older the rock, the steeper the slope. The intercept of the isochron with the ordinate gives a measure of the initial daughter. At the moment, this description of the isochron may seem a bit cryptic, but the method is really quite simple, as the next few paragraphs will make clear.

The trick to the isochron diagram is the normalization of both parent and daughter isotopes to a third isotope. The daughter isotope in a rock or mineral always consists of two parts: the amount that was there when the rock formed (the common or initial amount) and the amount added by radioactive decay of the parent isotope over time. Normalization consists of dividing both parent and daughter isotopes by a stable isotope of the same element as the daughter isotope. This stable isotope is not a product of radioactive decay, so it varies exactly with the amount of the initial (or common) daughter isotope. In other words, when the amount of one isotope goes up or down, the amount of the other isotope goes up or down in exactly the same proportion, so even though the amounts change, the ratio of the two isotopes never does. This happens because natural chemical and physical processes almost never fractionate isotopes of the same element with as little mass difference as there is between the isotopes used for radiometric dating. In order to see exactly what this normalization does and how the isochron works, consider what happens when the data are not normalized and consist solely of the amounts of the parent and daughter isotopes.

Suppose that two minerals, P and B, are separated from a rock of zero age, the amounts of ^{87}Rb and ^{87}Sr in the minerals and in a representative sample of the rock are measured, and this information is plotted on a graph (Figure 4.4a). Let's assume the results can be expressed in small numbers of atoms. If these samples sit for some time and are later reanalyzed, the points on the graph will have moved because of the decay of ^{87}Rb to ^{87}Sr (Figure 4.4b). In each sample, the decay of one atom of ^{87}Rb results in an increase of exactly one atom of ^{87}Sr, so the points will have moved along trajectories of 45° toward decreasing ^{87}Rb and increasing ^{87}Sr. Since the number of atoms of ^{87}Rb that decay in any time period is proportional to the number present, the distance the points move along their trajectories is a direct function of their ^{87}Rb contents. Thus, W will move two units to the left and two up because it has twice the ^{87}Rb content as P. Point B has seven times as much ^{87}Rb as P so it will move seven times farther along its trajectory than P.

The information parts a and b of Figure 4.4 provide is mimimal. If both the before and after amounts were known, an age for the rock could be calculated because the distance each point had moved along its trajectory is a function of time. But the original compositions of P, B, and W cannot be found, so there is no way of calculating the age. We're not stuck, though, because normalization provides an ingenious way around this problem.

See what happens when these same hypothetical data are normalized. To

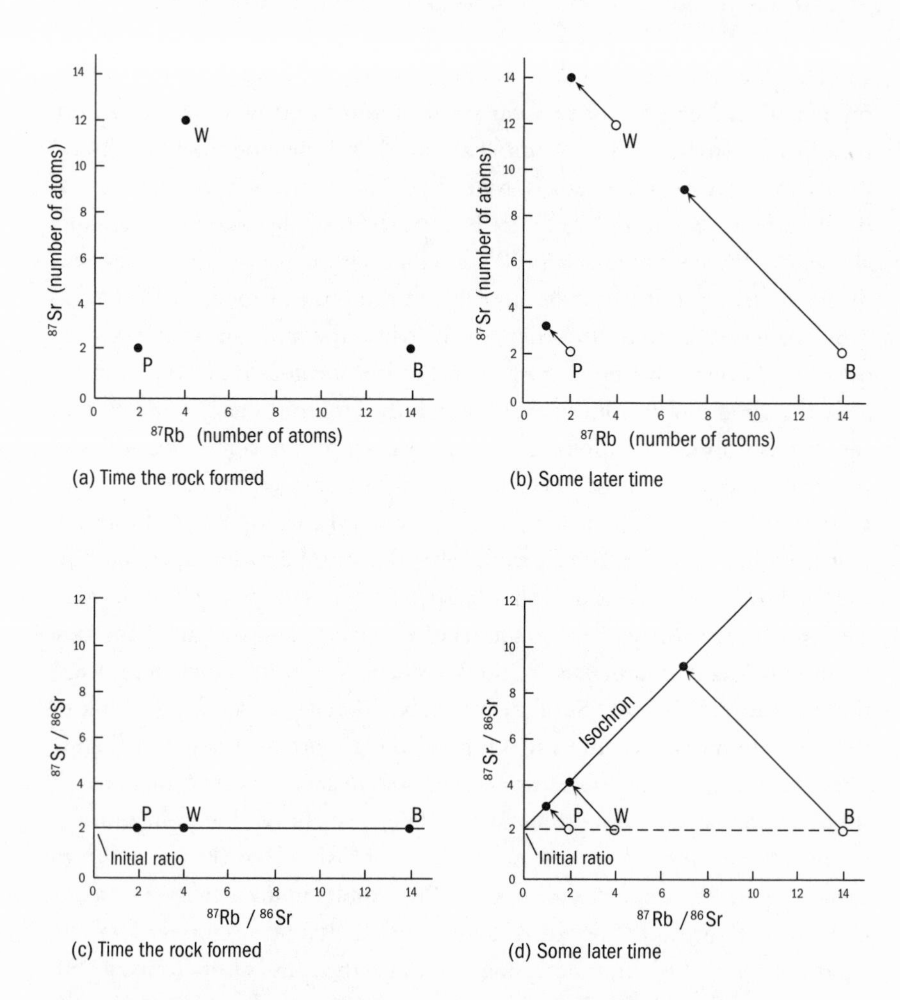

Figure 4.4 How isotope ratios reveal age. (a) A plot of ^{87}Rb vs. ^{87}Sr for two minerals, P and B, and the whole rock, W, from a hypothetical rock of zero age. (b) As ^{87}Rb decays, the points move along trajectories of decreasing ^{87}Rb and increasing ^{87}Sr (arrows). The movement is proportional to the ^{87}Rb content, but this type of plot gives no information about the age of the rock. (c) The same data but normalized to ^{86}Sr. (d) As time passes, the data will fall on a line, or isochron, whose slope is a function of age. In such a graph, called an isochron diagram, the intersection of the isochron with the ordinate gives the ^{87}Sr/^{86}Sr ratio of the rock when it first formed.

normalize, the amounts of both the ^{87}Rb and ^{87}Sr are divided by the amount of ^{86}Sr, which is neither radioactive nor the product of any decay. Let's say that for every two atoms of ^{87}Sr in sample P there is one atom of ^{86}Sr. This means that both ratios $^{87}Sr/^{86}Sr$ and $^{87}Rb/^{86}Sr$ for sample P are equal to 2. If the ratio $^{87}Sr/^{86}Sr$ is 2 for sample P, then it must also be 2 for samples B and W, so the data plot on a straight, horizontal line with $^{87}Sr/^{86}Sr$ equal to 2 for all values of $^{87}Rb/^{86}Sr$ in the three samples (Figure 4.4c).

Why are the strontium isotope ratios the same even though the rubidium and strontium contents of the minerals vary? The reason is that when a rock forms, all of the isotopes of each element are homogenized. Even though the amount of strontium varies from mineral to mineral, the chemical processes of crystallization do not fractionate isotopes of the same element, as long as their atomic masses are greater than about 20. As a result, the isotope composition of the strontium (and the rubidium) is the same in all the minerals and in the rock as a whole. Thus, for a rock whose age is zero, the ratios of $^{87}Sr/^{86}Sr$ in all of the minerals will be identical, whereas the ratios $^{87}Rb/^{86}Sr$ will vary from one mineral species to another, as in Figure 4.4c. The isochron for this zero-age rock is a horizontal line.

As our hypothetical rock ages, the isotope compositions of the samples P, B, and W will move along their respective trajectories as before and for the same reasons (Figure 4.4d). Each decay of ^{87}Rb results in the addition of an atom of ^{87}Sr, so as the ratio $^{87}Rb/^{86}Sr$ decreases, there is a corresponding increase in $^{87}Sr/^{86}Sr$, and the magnitude of the change is a function of the rubidium content, or in this case of the ratio $^{87}Rb/^{86}Sr$. With the normalized data, however, the isotope ratios will always fall on a straight line whose slope is a direct function of the age of the rock. The older the rock, the steeper the slope of the isochron. Furthermore, the isochron will always intersect the ordinate at the value of the initial isotope composition of strontium. This must be so because the ratio $^{87}Rb/^{86}Sr$ is zero at this intersect, and so there can be no increase in ^{87}Sr over time. The isochron pivots about this fixed point as the rock ages. Thus, the isochron method gives both the age and the isotope composition of the initial daughter solely on the basis of present-day isotope measurements.

Next imagine what happens to the isochron when the Rb-Sr clock is reset. Suppose the rock is completely melted and allowed to recrystallize so that it becomes a new rock with a new life. Melting will rehomogenize the strontium isotopes and the new minerals will once again share the same ini-

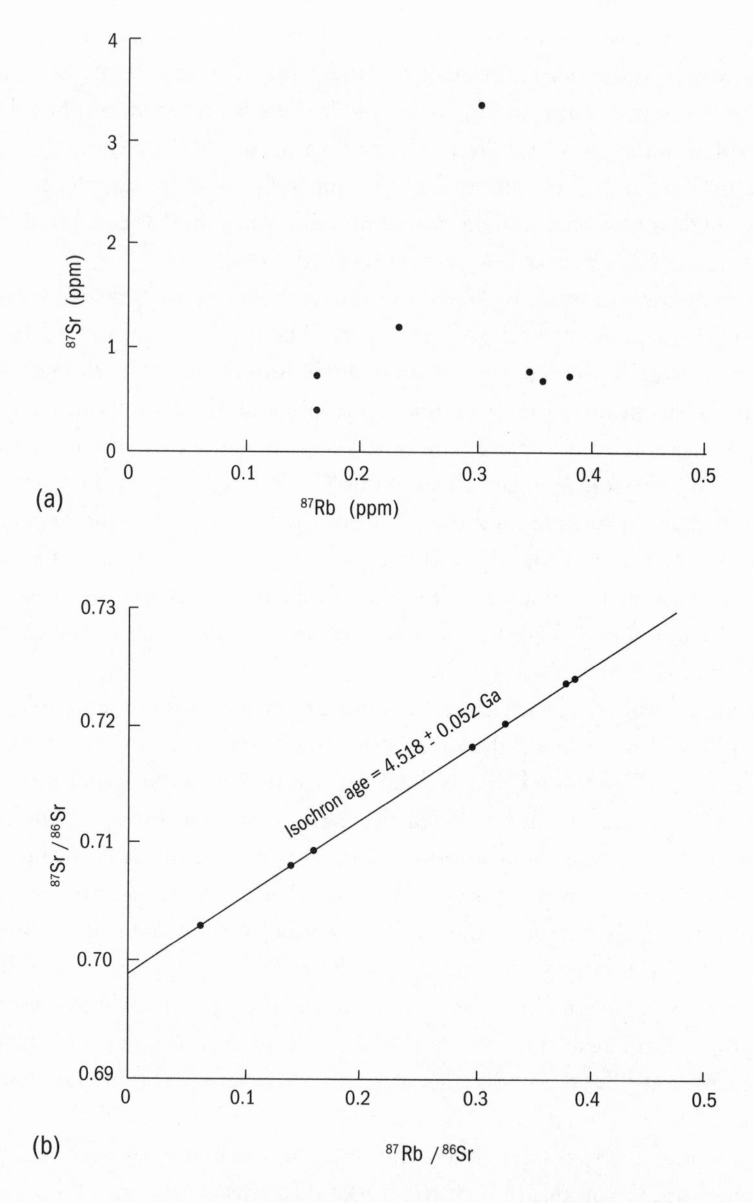

Figure 4.5 Rubidium and strontium in minerals and whole-rock samples of the Tieschitz meteorite. (a) Amounts of ^{87}Rb and ^{87}Sr in ppm. (b) Isotope ratios. Graph (a) reveals little information of interest, whereas graph (b) shows a linear relationship that not only provides the isochron age of 4.518 ± 0.052 Ga but also demonstrates that neither rubidium nor strontium has been added or lost since Tieschitz formed. (Data from Minster and Allègre 1979.)

tial ratio of $^{87}Sr/^{86}Sr$, but this ratio now will be higher than it was the first time around because of the decay of ^{87}Rb during the rock's first life. Graphically, it is as if the isochron rotates clockwise about the isotope composition of the total rock (point W, solid circle, in Figure 4.4d) and becomes horizontal again at the new initial ratio. The remelting and recrystallization will have completely reset the Rb-Sr clock. A later age measurement will reflect the most recent time of recrystallization and the most recent initial Sr isotope composition; all indications of the previous age and initial composition will have been erased.

If the rock is heated insufficiently to completely rehomogenize the strontium isotopes but enough so that strontium or rubidium is allowed to move about, then the clock may not be completely reset. In such a disturbed system either strontium isotopes or rubidium isotopes, or both, may move in or out of the rock, or the isotopes may simply be redistributed among the different minerals but not homogenized. In such instances the data will not fall on an isochron; they will scatter. The scattering will tell us that the sample has been disturbed and is undatable with this technique.

An example of some actual Rb-Sr isochron data is shown in Figure 4.5. Figure 4.5a shows the data for the meteorite Tieschitz plotted as parent and daughter isotope amounts. As with the prior example of female populations, there is nothing wrong with this graph, but it doesn't provide any useful age information. When the Tieschitz data are plotted as isotope ratios, however, a beautiful isochron emerges (Figure 4.5b). For the Tieschitz meteorite, the only reasonable conclusion is that its age is 4.52 Ga. For the data to fall on an isochron for any reason other than the decay of ^{87}Rb within a closed system over time would be a highly improbable coincidence.

The isochron diagram is used extensively for the Sm-Nd, Lu-Hf, and Re-Os methods in exactly the same way that it is used for Rb-Sr dating. The only difference is the use of other isotope ratios on the ordinate and abscissa.

THE AR-AR AGE SPECTRUM

The Ar-Ar method is just K-Ar dating with a twist. The sample is irradiated with energetic neutrons in a nuclear reactor in order to convert a fraction of the ^{39}K, the most common isotope of potassium, to ^{39}Ar. The irradiation results in the addition of a neutron and the ejection of a proton, which changes the ^{39}K to ^{39}Ar. Instead of measuring the amounts of potassium and argon separately by different methods, as is done in the conventional K-Ar method, it is possible to determine the exact ratio of daughter to parent in the sample by measuring the ratio of ^{40}Ar to ^{39}Ar. Even though the ^{39}Ar is produced in the reactor from ^{39}K, it serves as a proxy for ^{40}K because the ratio of $^{40}K/^{39}K$ is the same today in all matter, organic and inorganic. Corrections must be made for the efficiency of the conversion, for the atmospheric argon present, and for certain interfering nuclear reactions, but these can be made quite precisely and for old rocks they are usually small or negligible. When all of the argon in a sample is measured at the same time, this method yields a simple accumulation age equivalent to one determined by the conventional K-Ar method, except that it's usually more precise because isotope ratios can be measured better than amounts.

The Ar-Ar technique has some advantages over the conventional K-Ar method, including increased precision and smaller sample size requirements. (With the new laser systems and very sensitive mass spectrometers now in use, samples weighing as little as a millionth of a gram can be dated.) The primary advantage of this technique, however, is the ability to heat the sample incrementally to progressively higher temperatures, and collect and analyze the argon isotopes released at each temperature. An age is then calculated for each gas increment.

The series of ages from an incremental heating experiment are plotted against the percentage of ^{39}Ar released. This type of diagram is called an *age spectrum*. For an ideal, undisturbed sample, the calculated ages for the successive gas increments are all the same, and the age spectrum is a horizontal line, or band when analytical uncertainties are included, at the value corresponding to the age of the rock. These same data can also be plotted on a Ar-Ar isochron diagram, and they will fall on a straight line whose slope is a function of age and whose intercept is the $^{40}Ar/^{36}Ar$ ratio of atmospheric argon. The only difference between the age spectrum and isochron diagrams is that the latter does not require any assumption about the composition of nonradiogenic argon; otherwise, the two diagrams are just two methods of visually displaying the same data and almost always give the same age.

For a sample that is heated at some time after formation, the calculated ages for the individual gas increments may not all be the same. The increments released at the lower temperatures are usually disturbed, because they represent argon released from near mineral grain boundaries. Argon is most easily released from these sites by heating in nature. The high-temperature

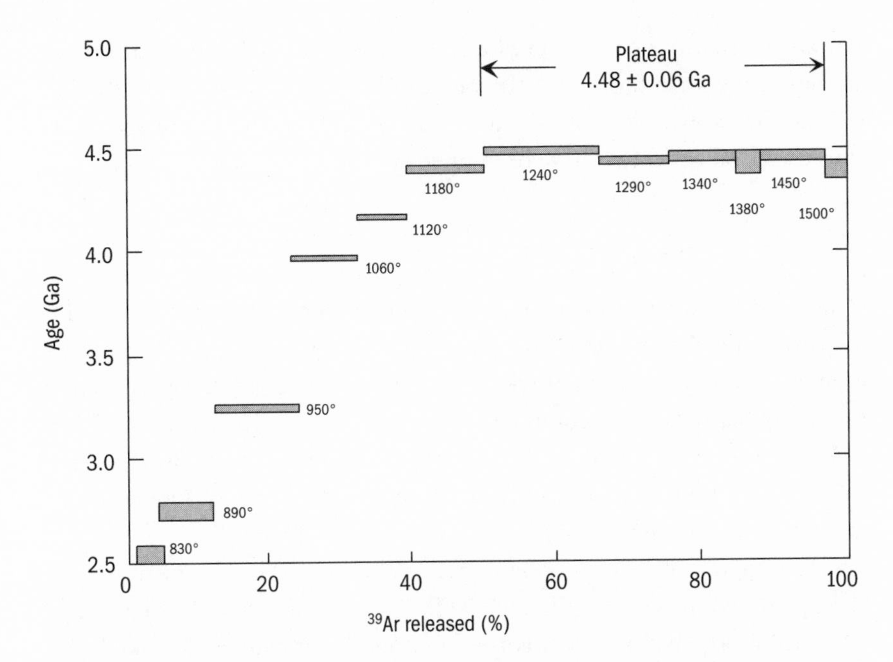

Figure 4.6 An Ar-Ar age spectrum for the Menow meteorite. Menow lost 25% of its Ar at about 2.5 Ga. The temperature at which each gas increment was released is shown in degrees Celsius. The vertical height of each box indicates the uncertainty in the measurement. The plateau in the age spectrum gives the age of Menow even though the meteorite lost some argon several billion years after it formed. (After Turner et al. 1978.)

increments may not be affected and may form a "plateau" that still reflects the original crystallization age of the rock. The plateau increments will still fall on a meaningful isochron, whereas the disturbed increments will not. Disturbed samples may give more complicated patterns, some of which are uninterpretable, but a high-temperature plateau age can usually be safely interpreted as the formation age of the rock or mineral.

An age spectrum for the chondrite meteorite Menow is shown in Figure 4.6. From the shape of the age spectrum, it is possible to calculate that approximately 25% of the total ^{40}Ar was lost, perhaps due to heating induced by collision with another body, at about 2.5 Ga. The high-temperature increments, however, still reflect the age of Menow—an age confirmed by Rb-Sr dating. The Ar-Ar isochron age from this age spectrum is also 4.48 Ga.

The data from an Ar-Ar incremental heating experiment are, like the Rb-Sr isochron method, self-checking. Departure of the data from a horizontal line and scatter in the isochron plot reveal that a sample has been disturbed at some time after formation. If the sample has not been too badly disturbed, such as with Menow, the method may still provide a valid age and is thus useful on rocks that have been open systems.

U-Pb concordia and discordia

One of the most powerful and reliable dating methods available is the U-Pb concordia method. Like the Ar-Ar age spectrum, this method can be used even when the rock has been reheated or altered, a feature of special value for dating old rocks with complex histories. The technique is self-checking. The U-Pb concordia method differs from the simple isochron method in a fundamental way because it uses the simultaneous decay of two uranium isotopes to two isotopes of lead: ^{238}U to ^{206}Pb and ^{235}U to ^{207}Pb.

If a mineral or rock initially contains ^{238}U and ^{235}U but no lead, the change in the ratios $^{206}Pb/^{238}U$ and $^{207}Pb/^{235}U$ over time will define a single curve called *concordia*, which is the locus of all concordant (equal) U-Pb ages (Figure 4.7). All minerals that formed at the same time and have remained closed systems will plot at the same time-point on concordia and will also have identical $^{206}Pb/^{238}U$, $^{207}Pb/^{235}U$, and $^{206}Pb/^{207}Pb$ ages. Concordia is curved because ^{238}U and ^{235}U decay at different rates (see Table 4.1), and, therefore, the relative rates of production of the two lead isotopes have changed with the passage of time.

The graphing of concordant U-Pb data on a concordia diagram does not provide any information that is not obvious from the numerical U-Pb ages. The principal value of the concordia diagram is its ability to yield crystallization ages from disturbed rocks.

Lead is a volatile element and can be lost from minerals when they are heated. Lead loss, however, does not fractionate the lead isotopes, because they are chemically identical and their masses are very nearly the same. Thus, lead loss from a mineral with a composition on concordia will result in the ratios $^{206}Pb/^{238}U$ and $^{207}Pb/^{235}U$ changing along a straight line to the concordia curve, connecting the original age on concordia and the time on concordia of the lead loss. This line is called *discordia*.

As may be obvious, a discordia cannot be determined by a single point. There must be at least two and preferably three or more points along a discordia line before it is possible to find either the original age of the rock or

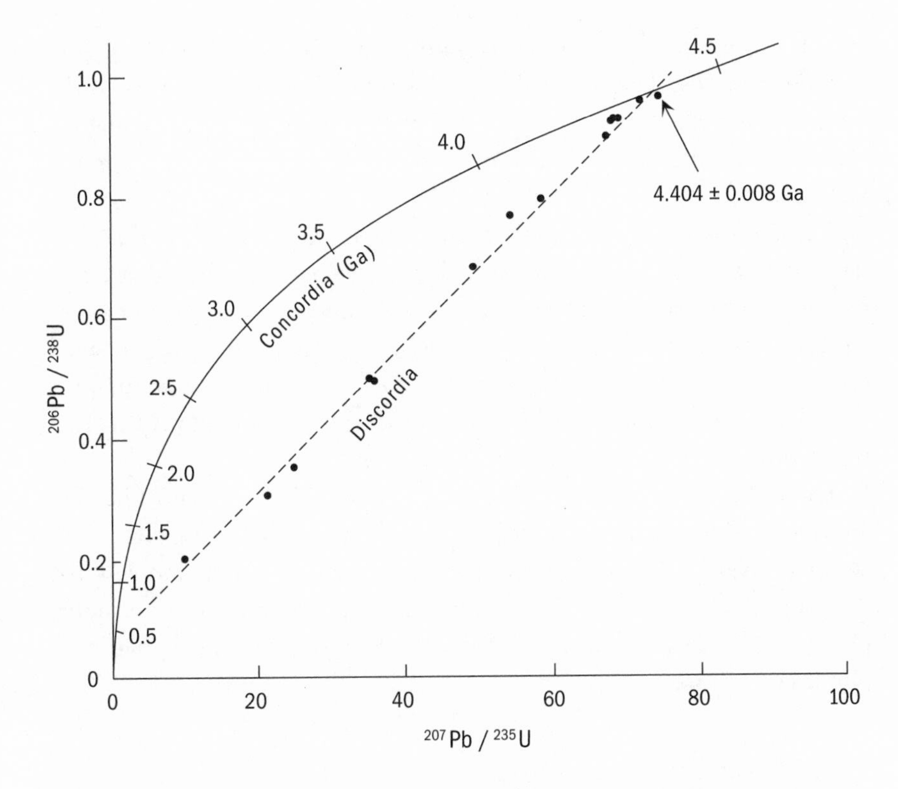

Figure 4.7 A U-Pb concordia diagram. This diagram shows the results of analyses on different parts of a zircon grain from the Jack Hills Conglomerate, western Australia, using an ion probe. The intersection of discordia (dashed line) with concordia gives the age of formation of the grain. The oldest part of this grain (arrow) very nearly falls on concordia and has a calculated Pb-Pb age of 4.404 ± 0.008 Ga. It is the oldest mineral yet found on Earth. (Data from Wilde and Valley 2001.)

the time of lead loss. As it turns out, however, this is not especially difficult. The amount of lead lost from a sample is controlled by a variety of factors, including grain size, original crystal imperfections, composition, and radiation damage from the decaying uranium. Because of this behavior, different crystals from the same rock sample, or even different parts of the same crystal, often will lose differing amounts of lead. Thus, analyses of different mineral grains from a rock sample, or even of different parts of the same mineral grain, will usually provide sufficient differences in uranium and lead isotope composition to define discordia. The example in Figure 4.7 shows analyses of different parts of a single zircon grain from a sedimentary formation in western Australia. The data fall on a discordia. The point farthest to the right falls very nearly on concordia and its Pb-Pb age is 4.404 Ga, making this tiny mineral grain the oldest object from Earth that has ever been found and dated.

The concordia-discordia method can be used only on minerals that contain either no initial lead or initial lead in such small quantities that a correction for its presence can be made accurately. Such minerals are not abundant in rocks, but fortunately several occur frequently, although in small amounts, in most igneous and metamorphic rocks. The most common is the mineral zircon, whose crystal structure accepts uranium but rejects lead so that initial lead is invariably negligible or very small.

The U-Pb concordia-discordia method is especially resistant to heating and metamorphic events and therefore is extremely useful in rocks with complex histories. Quite often this method is used in conjunction with the K-Ar and the Rb-Sr isochron methods to decipher the history of metamorphic rocks, because each of these methods responds differently to metamorphism and heating.

THE PB-PB ISOCHRON

Data from U-Pb systems can be treated solely in terms of the daughter isotopes of lead normalized to ^{204}Pb, which is neither radioactive nor a daughter isotope. This results in a Pb-Pb isochron, which passes through a point that represents the composition of the initial lead in the system. As a rock grows older, its lead isochron "rotates" about this initial composition to decreasing slopes. Unlike the simple isochron method, it is not possible to determine the initial lead isotope composition from the isochron data because it is not defined by the intersection of the isochron with one of the coordinates of the graph. The initial composition always lies somewhere along the isochron to the left of the measured data, but it is not possible to tell where. Nevertheless, the slope of the isochron alone reveals the age, so the method can be used to determine the age of a system without knowing either the composition or the amount of initial lead. Like the other isochron methods, this one is self-checking, and scatter indicates unreliable data that is most likely due to a geologic disturbance of some sort. An example of a Pb-Pb

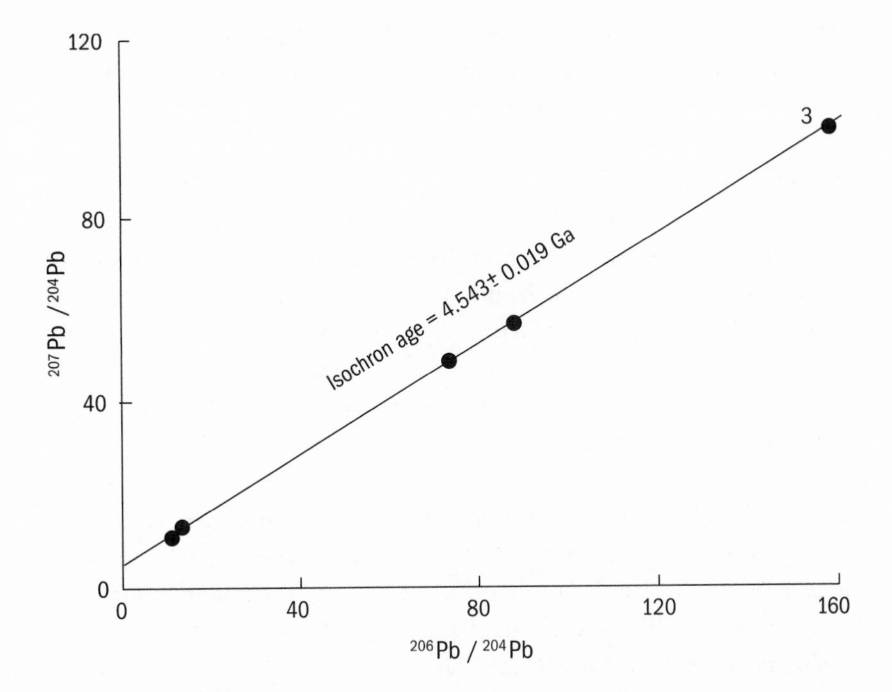

Figure 4.8 A Pb-Pb isochron for five samples from the Saint Severin meteorite. (After Manhès et al. 1978.)

isochron is the data for the meteorite Saint Severin, a former traveler of the Solar System that is 4.54 Ga old (Figure 4.8).

The Accuracy of Radiometric Dating

In the previous discussion of radiometric methods, the rationale for concluding that radioactive decay is constant over time, the geologic factors that can partially or totally reset radiometric clocks, and the various ways in which the graphical methods are self-checking were all described. But a few other factors deserve some attention.

First, all the half-lives listed in Table 4.1 have been determined by direct laboratory counting experiments and are known to an accuracy of about 2% or, for most parent isotopes, better. The uranium decay constants, in particular, are known to an accuracy of a small fraction of a percent. The uncertainties in decay constants are significant only when attempting to distinguish between the ages of early events in the Solar System using different dating methods; they do not significantly affect the values for the ages of Earth, the Moon, or meteorites. It is also worth noting that scientists worldwide use the same decay constants and isotope compositions for their calculations so that results from different laboratories can be easily compared.

Second, the modern analytical instruments used for isotope measurements (mass spectrometers of various types, including ion probes) are so reliable that the precision of the laboratory measurements usually exceeds the errors introduced by geologic factors. Most isotope ratios can now be measured to an accuracy of a few tenths of a percent or better.

Finally, the uncertainties in radiometric ages, although mathematically determined, highly useful, and generally realistic, are statistical estimates and do not define precise limits within which the "real" age must fall. These "errors" include only analytical uncertainties, not unrecognized geologic factors that might introduce additional errors in the measured age. The uncertainties assigned to the ages in this book are of analytical precision at the 95% confidence level, which means that the odds are 19 out of 20 that another measurement would fall within the stated error.

So, does all of this mean that all radiometric ages are right? Not quite. Literally tens of thousands of radiometric ages have been measured in many laboratories around the world, and it would be miraculous if some of them were not wrong. The methods and the people who use them are not infallible. Those of us who develop and use dating techniques to solve scientific problems are well aware that the systems are not perfect, and we have provided numerous published examples in which the techniques fail. We often test them under controlled conditions to learn when and why they fail so we can avoid using them incorrectly. Methods that have proven unreliable have been abandoned. For example, after extensive testing over many years, it was concluded that U-He dating is unreliable because the small helium atoms diffuse too easily out of minerals over geologic time. Much like a broken watch, the U-He method rarely gives the right time even though it continues to tick. As a result, this method is no longer used except in special applications. Other dating techniques, like those listed in Table 4.1, have stood the test of time. These methods provide valid age data in most instances, although there is a small percentage of instances in which even these generally reliable methods yield incorrect results. Such failures may be due to laboratory errors (mistakes happen), unrecognized geologic factors (nature sometimes fools us), or misapplication of the techniques (no one is perfect).

We scientists who measure isotope ages do not rely entirely on the error estimates and the self-checking features of age diagnostic diagrams to evaluate the accuracy of radiometric ages. Whenever possible we design an age study to take advantage of other ways of checking the reliability of the age measurements. The simplest means is to repeat the analytical measurements in order to check for laboratory errors. Another method is to make age measurements on several samples from the same rock unit. This technique helps identify post-formation geologic disturbances because different minerals respond differently to heating and chemical changes. The isochron techniques are partly based on this principle.

The use of different dating methods on the same rock is an excellent way to check the accuracy of age results. If two or more radiometric clocks based on different elements and running at different rates give the same age, that's powerful evidence that the ages are probably correct.

Geologic relationships also constitute an important way to evaluate radiometric ages. For example, a series of age measurements on rock bodies whose relative ages are known from their stratigraphic positions should fall in the same relative sequence. If they don't, then something is amiss with either the radiometric data or the interpretation of the field observations.

The next five chapters contain samples of the radiometric data relevant to the ages of Earth, the Moon, and meteorites. Examples of the various ways in which the age measurements are confirmed are included.

CHAPTER FIVE

Earth's Most Ancient Rocks

If you stand on a forested ridge and survey your immediate surroundings, the shape, color, and structure of the leaves, the tree bark, and the forest floor are quite clear. As you attempt to see farther and farther away, however, the details become progressively less and less sharp, and on the horizon only a hazy ridgeline is discernible. Binoculars help, but only a little—the distant view is still vague. The situation is much the same for scientists trying to decipher Earth's history. We know much about the planet's most recent past, but only a small fraction of what we would like to know about its earliest history. The details become dimmer and dimmer as we peer ever deeper back in time.

The Precambrian represents 87% of geologic time (see Figure 3.1). Although Precambrian rocks underlie more than half of the present-day land area, we know far less about Precambrian geologic history than about Phanerozoic history. There are several reasons for this. First and foremost is the rarity of fossils in Precambrian rocks. Precambrian life lacked the hard skeletal parts necessary for preservation as fossils except under extraordinary conditions. It also lacked the abundance and diversity required for use as distinctive stratigraphic age indicators. From the time of the earliest known fossils 3.5 billion years ago until more complex life began to develop in the latest Proterozoic, life on Earth consisted of single-celled organisms that lacked hard parts and diverse features. So the use of distinctive differences in fossils, one of the primary tools used for determining the relative sequence of rock units and geologic events in the Phanerozoic, is largely unavailable for use in Precambrian studies.

Second, many of the older Precambrian rocks have been highly altered by one or more episodes of metamorphism and deformation. As a result, they tend to be highly deformed and the relative age relations between rock units, as well as the original rock types, are often obscure or decipherable only with difficulty and uncertainty (Figure 5.1).

Third, the deposition of younger rocks and erosion have both concealed and removed Precambrian formations, so the accessible geologic record of Precambrian events, especially the earliest ones, is much less complete than that of Phanerozoic events. Despite these considerable difficulties, however, scientists are gradually unraveling the story of Earth's earliest history, and quite a bit is known about the Precambrian Earth.

Precambrian rocks are common and occur on all continents. Rocks of Proterozoic age (0.57–2.5 Ga) are the most widespread. They occur over more than two-thirds of North and South America, Africa, India, and Greenland and are common over large areas of Australia, Europe, Asia, and Antarctica. Archean rocks (2.5–4.0 Ga) also occur on all continents but over small and roughly equidimensional areas called cratons, blocks, or shields. The typical Archean craton is only about 0.25–0.5 million km^2 in area, but it is likely that Archean rocks are hidden beneath many areas of younger rocks, and Archean rocks may constitute nearly half or more of Earth's present-day continental mass. Terrestrial Priscoan rocks (>4.0 Ga) have been found in only one small area of northern Canada. Minerals of Priscoan age, however, are known from Australia.

Although Priscoan rocks are extremely rare, rocks with ages of 3.5–3.9 Ga are found on nearly all continents (Table 5.1). Rather than describe every occurrence of early Archean rocks and the evidence for their age, I will briefly describe the geologic setting and evidence for the ages of early Archean rocks in three areas that have been well studied: the North-Atlantic craton of Greenland and Labrador, the Slave craton in northern Canada, and the Pilbara block of western Australia (Figure 5.2). Before plunging into the geology of these three areas, however, a few general remarks about rocks in general and Archean rocks in particular may be helpful.

A Word About Rocks and Archean Rocks

There are literally hundreds of names for rocks, and it is a rare geologist who knows them all. There are also several ways of classifying and naming rocks, according to their chemistry, minerals, texture, and origin. Rock names are a valuable form of geologic shorthand, conveying in a single word, some-

Figure 5.1 The Acasta Gneiss, near Slave Lake in the Northwest Territories of Canada, the oldest Earth rock yet found. U-Pb dating shows that this rock first crystallized 4.03 billion years ago. Originally a granitoid intrusion, the rock is now altered and deformed by metamorphism. The darker bands are concentrations of mafic minerals. The pocket knife is about 8 cm (3 inches) long. (Photo courtesy of S.A. Bowring.)

Table 5.1

Locations, Types, and Ages of Some of Earth's Oldest Known Rocks

Location	Name	Dominant Rock Types: Now	Dominant Rock Types: Original	Dating Method	Age (Ga)
Africa, Swaziland	Ancient Gneiss Complex	Tonalite gneiss	Granitoid	Pb-Pb, U-Pb	3.64
Antarctica, Mt. Sones	Napier Complex	Tonalite gneiss	Granitoid	Pb-Pb, U-Pb	3.93
Australia, western	Narryer Gneiss Complex	Tonalite gneiss	Granitoid	Pb-Pb, U-Pb	3.73
	Northstar Basalt	Ultramafic to felsic lava flows	Lava flows	Sm-Nd	3.56
Canada, Northwest Territories	Acasta Gneiss Complex	Tonalite to granodiorite gneiss	Granitoids	Pb-Pb, U-Pb	4.03
China, northeast	Anshan Complex	Trondhjemite gneiss	Granitoid	U-Pb	3.81
Greenland, western	Itsaq Gneiss Complex	Gneiss, supracrustals	Granitoids, sedimentary rocks, lava flows	U-Pb, Pb-Pb, Rb-Sr, Sm-Nd	3.57–3.75
Labrador, northern	Archean Gneiss Complex	Ultramafic rocks	Ultramafic intrusions, lava flows	Sm-Nd	3.82–4.02
Ukraine	Novopavlovsk Complex	Ultramafic rocks	Ultramafic intrusions	Pb-Pb, U-Pb	3.64
United States, Minnesota R. Valley	Morton Gneiss, gneiss near Granite Falls	Granitoid gneisses	Granitoids	U-Pb, Rb-Sr	3.48–3.68
Venezuela	Imataca Gneiss Complex	Granitoid gneisses	Sedimentary rocks	Pb-Pb	3.77
Zimbabwe	Sand River Gneisses	Diorite to granodiorite gneisses	Granitoids	Rb-Sr	3.73–3.79

NOTES: Only rocks with radiometric ages near or exceeding 3.6 Ga are listed. There are a great many more rocks with ages of 3.5 Ga or less.

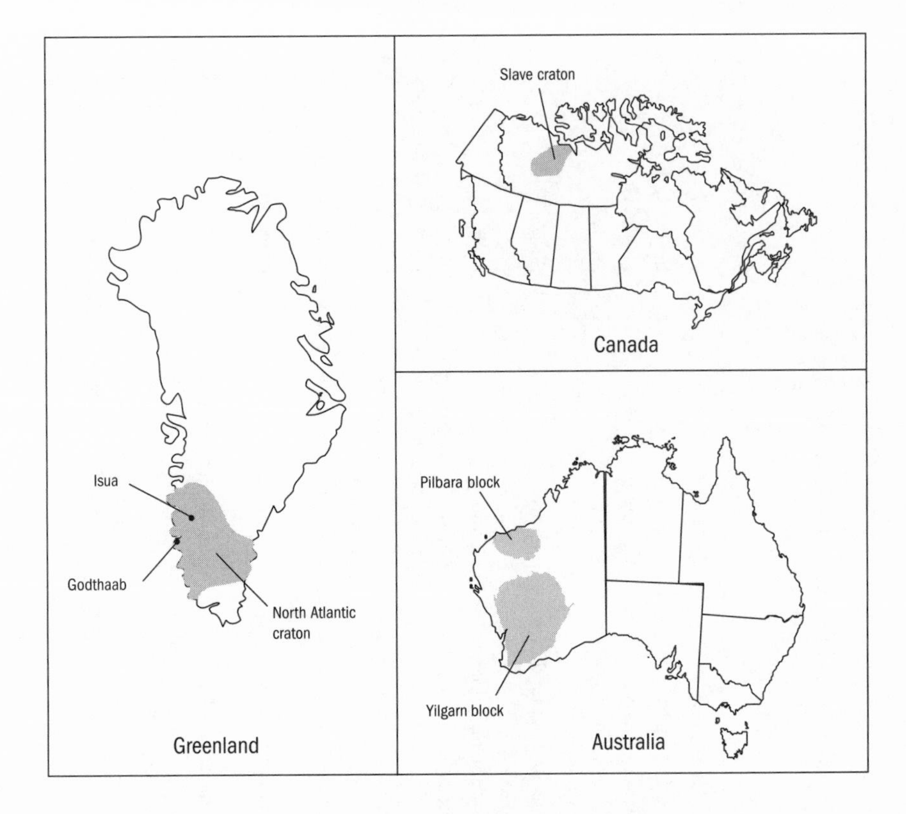

Figure 5.2 Locations of the ancient rocks described in the text.

times two, much about the composition, texture, and origin of a rock. For our purposes, however, we need only a few simple terms.

There are three basic types of rocks: igneous, sedimentary, and metamorphic. *Igneous rocks* crystallize from molten rock, called magma. If the magma cools and crystallizes below Earth's surface, the rocks are called plutonic rocks; if the magma erupts onto the surface as lava, the resulting rocks are called volcanic rocks. Igneous rocks are generally classified by their compositions, with different names for the volcanic and plutonic equivalents (Table 5.2).

Sedimentary rocks are of two general types, detrital and chemical. Detrital rocks are composed of pre-existing mineral and rock grains. They form as rocks weather and erode, and wind or water redeposit the debris. Sedimentary rocks are usually named by their grain size. Shale is composed of clay-sized (microscopic) particles, and sandstone is composed of sand-sized grains. Conglomerate contains boulders, cobbles, and pebbles embedded in a matrix of

Table 5.2

A Simplified Classification of Some Important Igneous Rocks

General Mineralogical Groupings	Volcanic Rocks	Plutonic Rocks		Chemical Composition
Felsic	Rhyolite	Granite	Granitoids	Silicon ↑, Sodium, Potassium ↑, Calcium ↓, Iron, Magnesium ↓
		Monzonite		
	Dacite	Granodiorite		
Mafic	Andesite	Diorite, tonalite		
	Basalt	Gabbro		
Ultramafic	Komatiite	Periodotite		

NOTE: Arrows indicate increasing concentrations of some key elements.

sand and clay. Most detrital sedimentary rocks are deposited in oceans, lakes, and rivers, although some are deposited by wind (sand dunes) and others by glaciers (glacial deposits). The most common type of chemical sedimentary rock is limestone, which is chemically precipitated from sea water. The deposition of the shells of floating, microscopic animals also forms limestone. Other common types of chemical sedimentary rocks are chert, a precipitate of silicon dioxide similar to quartz, and banded iron formation, which is a precipitate of iron-rich chert.

Metamorphic rocks are formed from igneous, sedimentary, or other metamorphic rocks by processes that occur deep in the crust. These rocks recrystallize and are deformed through the actions of heat, pressure, and chemical fluids, but they do not completely melt to form new magma. Many metamorphic rocks have been through more than one episode of metamorphism and deformation. Often a geologist can tell the original rock type from which a particular metamorphic rock formed, but sometimes the changes caused by the metamorphism have been so drastic that it is impossible. Metamorphic rocks are usually named by their textures, and there are two important types. Gneisses are coarsely crystalline, whereas schists are fine-grained and usually contain abundant mica. A gneiss may include thin bands of schist. A compositional term is often added to indicate the type of gneiss. Thus, granite gneiss has the composition of granite.

Metamorphic rocks predominate in the Precambrian, especially in the Archean eon. The substantial changes that have occurred over time in older rocks by metamorphism and the lack of fossils in Precambrian rocks are the primary reasons that Precambrian rocks are so difficult to decipher and why Precambrian geologic history is not known in the same detail as is Phanerozoic geologic history.

A typical early Archean terrane can be subdivided into three categories of rock units, including gneisses, intrusive rocks, and supracrustal rocks. About 80–90% of Archean rocks are gneisses, which occur as vast "seas" that surround, intrude, and form the basement for both older and younger rocks. Most were injected as granitoids and were changed to gneisses by high pressure and temperature deep within the ancient continental crust; they probably represent new additions to a growing Archean continental crust. The process may be similar to the one that generated the Mesozoic granitoid batholiths, which form the cores of the great mountain ranges of the west coasts of North and South America: the Sierra Nevada and the Andes. If the rocks that form the hidden roots of these magnificent mountain ranges could be examined, it is likely that gneisses, not granitoids, would be found.

Archean supracrustal rocks are rocks that were deposited on pre-existing crust. They include sandstones, shales, conglomerates containing boulders and pebbles of still older granitoids and volcanic rocks, cherts, limestones, and, occasionally, banded iron formation. The lava flows include both mafic and ultramafic types, although the most common ones resemble basalt similar to the lavas that erupt from volcanoes in Hawaii and on the seafloors. Some of the lava flows are still identifiable as pillow lavas, named for the pillow-shaped structures that form when lava flows and solidifies under water. These supracrustal rocks sometimes occur in accumulations that may reach a thickness of several kilometers (Figure 5.3). Archean supracrustal rocks were deposited in shallow inland basins or seas on the edges of the ancient continents, and most have been subjected to low-grade metamorphism.

There is nothing especially unusual about Archean supracrustal rocks, and all but one have recent analogs. Only banded iron formation, which is common in the early Precambrian and is an important source of iron worldwide, does not occur in the late Proterozoic and Phanerozoic. These are chemically precipitated sedimentary rocks that consist of alternating layers of iron-rich and iron-poor chert. Many scientists think that these remarkable deposits, which are important sources of iron ore, precipitated as prim-

Figure 5.3 Archean supracrustal rocks. This LANDSAT photograph of Precambrian terrain in the northern part of western Australia, known as the Pilbara block, shows Archean granitoids (lighter areas) intruding supracrustal rocks that include older, metamorphosed, and highly deformed sedimentary and volcanic rocks (darker areas between the granitoids). U-Pb dating of the granitoids and of the volcanic rocks shows that they are more than 3.4 Ga in age. The Shaw Batholith is discussed in the text. The distance across the image is about 165 km. (LANDSAT scene ID 84008501301X0, reproduced with permission of Earth Observation Satellite Company, Lanham, MD.)

itive organisms evolved and began to flourish in the early Precambrian oceans. According to this idea, these single-celled organisms provided oxygen that combined with iron, which originated from volcanic eruptions and from the weathering of volcanic rocks, dissolved in sea water. Other scientists think an inorganic origin for these iron deposits is more likely. But whether organic or inorganic, the deposition of banded iron formation ceased forever when the overabundance of dissolved iron in Earth's ancestral oceans was exhausted.

Younger igneous rocks commonly intrude the older gneisses and supracrustal rocks. These younger rocks range in composition from mafic to felsic and in form from large bodies of batholithic size to dikes (tabular intrusions) only a few centimeters thick. The younger intrusive rocks are primarily of interest because, where dated, they provide a younger limit on the age of the older rocks they intrude. Where these rocks intrude some units but not others, they act as time markers that are of value in determining the relative sequence of rock units and events.

The North Atlantic Craton

The North Atlantic Archean craton includes parts of eastern and western Greenland, part of the Labrador coast of Canada to the west of Greenland, and small sections of northwestern Scotland and northern Norway to the east of Greenland (see Figure 5.2). Once part of a single continent, this Archean terrane was torn into sections when North and South America were separated from Europe and Africa to form the Atlantic Ocean nearly 200 million years ago. The rocks of this craton are primarily granitoid gneisses, which constitute about 85% of the surface exposures. The remainder consists of metamorphosed supracrustal rocks with minor amounts of later intrusive rocks.

Most of the rocks of the North Atlantic craton are 2.6–3.0 Ga, but this ancient continental fragment also includes some of the oldest rocks found on Earth. Because of their antiquity, the parts of the craton that include these oldest rocks have been intensely studied for more than three decades by many scientists. Among those leading these efforts are Stephen Moorbath of Oxford University and Alan Nutman of the Australian National University. A variety of radiometric dating methods have been applied to these rocks, and the data show that the oldest are 3.7–3.8 Ga in age. Areas of particular interest are near Godthaab and Isua in southern West Greenland.

The area near Godthaab, the capital of Greenland, is dominated by gneiss complexes of two distinct ages (Table 5.3). The oldest is called the Itsaq Gneiss Complex, and the name is fitting because the word *itsaq* means "ancient thing" in Greenlandic. The Itsaq Gneiss Complex includes the Amîtsoq Gneisses and older supracrustal rocks. The supracrustal rocks are metamorphosed mafic lavas, mafic and ultramafic intrusive rocks, clastic sediments, and banded iron formation.

Table 5.3
Archean Rocks of Southern West Greenland

Rock Unit Name	Dating Methods	Range of Ages (billion years)
Mafic igneous dikes		
Qôrqut Granite	Pb-Pb, U-Pb, Rb-Sr	2.52–2.58
Nûk Gneisses	Pb-Pb, U-Pb, Rb-Sr	2.60–3.08
Layered anorthosites and granites	Pb-Pb, Rb-Sr, Ar-Ar	2.75–2.83
Malene Supracrustals		
Mafic igneous dikes		
Itasaq Gneiss Complex		
Amîtsoq Gneisses	Pb-Pb, U-Pb, Rb-Sr, Lu-Hf	3.55–3.74
Isua Greenstone Belt and other supracrustals	Pb-Pb, U-Pb, Rb-Sr, Sm-Nd	3.60–3.81

NOTES: The formations are listed in their known relative order, from younger to older, as determined by field studies. These ancient rocks have counterparts of the same ages in Labrador, which was once joined to Greenland and separated 200 million years ago by continental drift.

SOURCE: Nutman et al. 1996, Appel and Moorbath 1999; compilation in Dalrymple 1991.

Mafic igneous dikes intrude the Amîtsoq Gneisses. The dikes provide a valuable stratigraphic marker because they intrude the Amîtsoq Gneisses and its older inclusions but do not intrude younger rocks.

The Malene Supracrustals contain a high proportion of mafic volcanic rocks, including pillow lavas, as well as sediments that were mostly derived from volcanic rocks. The Malene Supracrustals are not cut by the mafic dikes and thus must be younger than the Amîtsoq Gneisses. The supracrustals are, however, intruded by layered bodies of anorthosite, an igneous rock composed almost entirely of plagioclase feldspar, which is a common, light-colored, sodium-calcium-aluminum silicate mineral.

The Nûk Gneisses intrude the Amîtsoq Gneisses, the mafic dikes, the Malene Supracrustals, and the layered igneous rocks. Despite their similarity in appearance and composition, the Nûk and Amîtsoq gneisses can be told apart by the presence or absence of mafic dikes, which are ubiquitous within the Amîtsoq Gneisses but do not intrude the Nûk Gneisses.

The youngest rocks in the Godthaab district are the Qôrqut Granite and some younger mafic igneous dikes. These younger dikes postdate most of the deformation undergone by older rocks and are only mildly metamorphosed.

Isua is a remote mountainous area 100 km northeast of Godthaab near the edge of the inland ice sheet. Most of the rocks exposed near Isua are gneisses that are equivalent to the Amîtsoq Gneisses in composition and stratigraphic position, and they are commonly given the same name. As near Godthaab, the Amîtsoq Gneisses near Isua are cut by mafic igneous dikes. Rocks younger than the dikes do not occur at Isua, and both the dikes and the gneisses are much less deformed and metamorphosed than they are near Godthaab. The importance of the Isua locality is that a thick and relatively undisturbed sequence of very ancient supracrustal rocks is exposed there. These remarkable rocks form an incomplete oval belt approximately 12 x 25 km, completely enclosed by and, in places, intruded by the Amîtsoq Gneisses. These supracrustal rocks, collectively called the Isua Greenstone Belt, are among the oldest known rocks on Earth. Equally important, these rocks show that volcanism and sedimentation were taking place in the early Archean seas adjacent to the primitive continents.

The Isua Greenstone Belt consists of metamorphic rocks whose ancestors include fine-grained detrital sediments, a conglomerate, cherts, banded iron formation, felsic, mafic and ultramafic intrusive rocks and lavas, including pillow lavas. These rocks, although deformed and metamorphosed, still retain many recognizable features that show their supracrustal origin. It is probable that these ancient supracrustal rocks are the equivalent of the supracrustal rocks enclosed in the Amîtsoq Gneisses near Godthaab.

Because of their antiquity, the early Archean rocks near Godthaab and Isua have been the objects of a great many radiometric age measurements by a number of independent researchers. The results of these efforts are summarized in Tables 5.3 and 5.4, where it is easy to see that the radiometric ages are consistent with the relative sequence of rock units as determined solely by field relationships. The Qôrqut Granite is 2.5–2.6 Ga and, where dated, the other rock units have increasingly older radiometric ages going downward in the sequence with the rocks of the Isua Greenstone Belt having the oldest radiometric ages of 3.6–3.8 Ga.

The ages of several of the units within the Isua supracrustal sequence have been determined using four different methods (see Table 5.4). The conglomerate contains boulders of felsic volcanic rock embedded in a fine-grained matrix of volcanic debris. A Rb-Sr isochron on five whole-rock samples of boulders and matrix from the conglomerate, and three samples of a related schist, gives an age of 3.66 Ga. More recent analysis of five additional samples of the conglomerate unit and related rocks gives a Rb-Sr isochron age of 3.71 Ga, in agreement with the earlier results.

Table 5.4
Some Radiometric Ages of Early Archean Supracrustal Rocks
Near Isua, Southern West Greenland

Unit	Method	Age (billion years)
Various units combined	U-Pb	3.81 ± 0.02
Conglomerate	U-Pb	3.77 ± 0.01
	Rb-Sr	3.66 ± 0.06
	Rb-Sr	3.71 ± 0.07
Sedimentary rocks	Sm-Nd	3.74 ± 0.06
Felsic volcanic rocks	Pb-Pb	3.81 ± 0.002
	Pb-Pb	3.81 ± 0.004
	Pb-Pb	3.71 ± 0.003
Felsic and mafic volcanic rocks	Sm-Nd	3.78 ± 0.05
Mafic igneous unit and conglomerate	Sm-Nd	3.75 ± 0.04
Banded iron formation	U-Pb, Pb-Pb	3.70 ± 0.07

NOTE: All ages are based on the isochron (Rb-Sr, Sm-Nd) or the concordia-discordia (U-Pb and Pb-Pb) method.

SOURCE: Moorbath, Whitehouse, and Kamber 1997; Nutman et al. 1996; compilation in Dalrymple 1991.

The Isua conglomerate unit has also been dated by the U-Pb concordia-discordia method using eight single zircon grains separated from two of the boulders in the conglomerate. The eight analyses fall on a discordia line indicating a crystallization age of 3.77 Ga. Most of these data are relatively close to the U-Pb concordia curve, which indicates that these particular Isua rocks have not been highly disturbed by metamorphism. Another study found that sixteen analyses of zircons from ten samples representing various volcanic and sedimentary units within the Isua supracrustal sequence define a discordia line and an age of 3.81 Ga.

As is common in metamorphic rocks, the U-Pb age from the zircons is a few percent older than the Rb-Sr ages on the whole rock samples. This is probably because the Rb-Sr and U-Pb methods are measuring slightly different events. If the usual interpretation of U-Pb zircon data is valid, then the age is probably near the time of crystallization of the original volcanic rock from which the conglomerate was derived. The Rb-Sr whole-rock isochron age, however, may represents the time of metamorphism of the Isua supracrustal unit. Thus, the two slightly different ages found by the two different methods are not necessarily inconsistent. In any event, they don't differ by much.

A mafic igneous unit occurs at several places within the Isua sequence. Field relations and chemical data indicate that it was probably intruded into the sedimentary sequence during metamorphism. Because of its composition, the unit is not amenable to U-Pb, Pb-Pb, or Rb-Sr dating, but it has been dated by the Sm-Nd technique. Eight samples of this unit and four of the conglomerate unit fall on a single Sm-Nd isochron with an age of 3.75 Ga. The validity of including data from formations with two different origins and histories in the same isochron calculation is debatable because it is doubtful that there is a simple genetic relationship between the two units. However, the fact that these data fall on a common isochron indicates that they were formed at about the same time from one or more mantle sources with the same initial neodymium isotope composition. It is uncertain, however, whether the Sm-Nd age represents the age of intrusion and crystallization of the mafic unit or the age of metamorphism of the Isua sequence.

The Pb-Pb ages of zircons from several felsic volcanic units have been measured recently using an ion probe to measure the U-Pb isotopes in individual zircon crystals. Two such analyses of multiple zircon crystals give ages of 3.81 Ga, while another series of analyses yields an age of 3.71 Ga. It has been proposed, but not proven, that the 3.81 Ga ages represent the ages of zircon crystals from older rocks that were incorporated into the volcanic rocks as they erupted, a process known to happen in modern volcanoes. If so, then they do not measure the age of the felsic volcanic rocks, but that of some older unit.

Other recent studies have provided a Sm-Nd age of 3.74 Ga for sedimentary units within the Isua Greenstone Belt, and an age of 3.78 Ga for felsic and mafic volcanic rocks.

Banded iron formation occurs throughout the Isua sequence but is most abundant at the extreme northeast end of the supracrustal arc. Here the iron-rich layers become sufficiently abundant that they constitute an ore body, partly buried by the inland ice sheet, with estimated reserves of some 2 billion tons. The banded iron formation has been dated by the Pb-Pb method using samples of iron-rich chert, the interlayered carbonate rock, and both iron oxide and silicate minerals separated from the iron formation. The isochron age is 3.70 Ga and is thought to represent the time of metamorphism, although it may record the depositional age.

Although it is uncertain whether the various radiometric ages of formations within the Isua supracrustal rock sequence represent ages of metamorphism, deposition, or crystallization of parent igneous material, the consistency is remarkable, especially in view of the variety of units dated and the variety of methods used. The conclusion is inescapable that the age of the Isua supracrustal rocks is at least 3.7 Ga. The age of the source material for the conglomerate, as represented by the U-Pb age on zircons from the volcanic boulders, approaches, and may exceed, 3.8 Ga. The U-Pb zircon discordia data for the mixture of supracrustal units suggests that these rocks are, indeed, about 3.8 Ga in age.

The Amîtsoq Gneisses have been dated by four methods applied to a large number of samples from a variety of localities near Isua and Godthaab, and the resulting isochron ages all fall within the range of 3.55–3.74 Ga (see Table 5.3). Several of the units that postdate the Amîtsoq Gneisses have been dated, and the results are consistent with their known relative ages. The layered anorthosite unit gives Rb-Sr, Pb-Pb, and Ar-Ar ages of 2.75–2.83 Ga. The Nûk Gneisses and associated rocks from a variety of localities from both inland and along the coast of southern West Greenland give Rb-Sr, Pb-Pb, and U-Pb zircon ages of 2.60 Ga–3.08 Ga. It's likely that the various rock bodies that have been included as Nûk Gneisses were emplaced by multiple intrusions over a long period of time, and that the range of ages found for the Nûk Gneisses represent real differences in age. Finally, the age of the Qôrqut Granite has been measured as about 2.5–2.6 Ga by Rb-Sr, Pb-Pb, and U-Pb methods.

The Archean rocks of the Labrador coast in the region near Saglek and Hebron are similar in nearly all respects to those of southern West Greenland and have similar radiometric ages. This is not surprising because the Saglek-Hebron area was very close to the Godthaab area before the breakup of the North Atlantic craton 200 million years ago. Thus, the fieldwork and radiometric ages on the rocks of Labrador independently confirm the antiquity and geologic history of the North Atlantic craton.

Acasta Gneisses of the Slave Craton

The oldest rock known on Earth, as of early 2002, is located in northernmost Canada, between Great Slave Lake and the Arctic Ocean's Coronation Gulf. Here, occupying an area of 190,000 km^2, is the Slave craton, which is composed of a highly deformed sequence of Proterozoic and Archean gneisses and supracrustal rocks (see Figure 5.2). Although these rocks have been slow to yield their secrets, a general outline of the stratigraphy has been determined by field mapping. About 40% of the rocks belong to a formation of supracrustal rocks that includes turbidites and lava flows of various compositions. Turbidites are formed by underwater debris flows and can be seen forming in today's oceans near land where slopes are steep, such as the undersea Monterey Canyon in northern California. Radiometric dating shows that the lava flows were erupted between 2.72 and 2.65 Ga. Granitoid plutons that are 2.58–2.67 Ga in age intrude the supracrustal rocks.

Rocks older than the supracrustals occur in the western part of the Slave craton. These include a number of granitoid intrusions and the Acasta Gneisses, which exceed 2.8 Ga in age. Within these older rocks are remnants of supracrustal rocks whose ages relative to the gneisses are unclear from field relationships. Radiometric dating in this area was begun in the early 1980s by S.A. Bowring, then at Washington University in St. Louis, and his colleagues. Their initial U-Pb dating studies showed that some of the Acasta Gneisses exceed 3.5 Ga in age, and the hunt was on.

In 1989, Bowring and his team analyzed twenty-seven zircon crystals from two samples of gneiss using the ion probe at the Australian National University. The Pb-Pb age of the oldest group of these crystals is 3962 ± 3 Ma, which is the age of crystallization of the original tonalite and granite intrusions from which the gneiss was formed. Subsequent studies yielded even older ages. In 1999, Bowring and his colleagues found three rock samples that have Pb-Pb zircon ages of 4002 ± 4, 4012 ± 6, and 4031 ± 3 Ma (see Figure 5.1). The composition of the Acasta Gneisses shows that as long ago as 4 Ga, Earth had developed continental crust similar to that we see today, though perhaps not as extensive. The search continues for still older rocks, but so far the Acasta Gneisses are the oldest rocks known.

Australia's Pilbara Block

The western part of the Australian continent consists of two Archean cratons surrounded mostly by deformed Proterozoic rocks (see Figure 5.2). The Yilgarn block, the southerly of the two, covers an area of nearly 650,000 km^2. It is the site of Earth's oldest minerals. The Archean rocks of the Pilbara block are exposed over an area of only about 56,000 km^2, but nearly half of the craton is covered on the south by Proterozoic sedimentary rocks, so its total area may exceed 100,000 km^2.

The Pilbara block consists of complex, domelike granitoid plutons surrounded by deformed belts of supracrustal rocks (see Figure 5.3). Field studies indicate that the supracrustal rocks were deposited on granitoid basement and then intruded by the granitoid domes. The supracrustal rocks were deposited in shallow seas and include pillow lavas, cherts, banded iron formation, and shallow-water detrital sediments. The exact ages of most of the supracrustal rocks are unknown, but the younger of them are intruded by, and are therefore older than, granitoids 2.9–3.3 Ga in age, while two of the oldest lava flow formations have radiometric ages of 3.5–3.6 Ga.

The North Star Basalt is a volcanic formation, 2 km thick, that is the oldest supracrustal formation in the Pilbara block. Its name is somewhat misleading because the unit contains not only basalt lava flows, but also a variety of rock types including felsic, mafic and ultramafic lava flows, as well as sedimentary rocks. The most precise age for the North Star Basalt was obtained by the Sm-Nd method. Data from six volcanic units that ranged in composition from ultramafic to felsic form a good isochron with a Sm-Nd age of 3.56 ± 0.03 Ga, which indicates the age of eruption of these flows.

Two other attempts to date the North Star Basalt have yielded less precise ages. Samples of four lava flows, including three dacites and one andesite, gave a Sm-Nd isochron age of 3.56 ± 0.54 Ga. The large error is not due to any imprecision in the isotope measurements, but rather to the sample similarity in samarium and neodymium isotope composition and the resulting lack of "spread" along the isochron. Rubidium and strontium measurements were made on these same four samples. The three dacites gave an isochron age of 3.57 ± 0.18 Ga, but the andesite sample did not fall on the isochron. It may seem curious that the andesite sample behaved as a closed system for samarium and neodymium but not for rubidium and strontium, but this is probably due to the higher resistance of the Sm-Nd isotope ratios to post-formation metamorphism.

The Duffer Formation, which is younger than the North Star Basalt, has been dated by the U-Pb concordia-discordia method using zircons from a dacite lava flow. The 11 data fit a discordia line that gives an age of 3.45 ± 0.02 Ga. Three attempts to date the Duffer Formation using the Rb-Sr method have yielded only poorly defined isochrons with imprecise ages in the range of 3.2–3.5 Ga.

The oldest granitoids found so far in the Pilbara block are from the Shaw Batholith (see Figure 5.3), from which U-Pb ages on zircons of 3.42 ± 0.04 and 3.47 ± 0.02 Ga have been obtained. Since the batholith intrudes the supracrustal rocks, it must be younger, so the radiometric ages are consistent with the field relations.

Other Ancient Rocks

The Superior Province of North America is Earth's largest known Archean crustal block and is mostly 2.6–2.8 Ga in age. A small area immediately to the south of Lake Superior, however, consists of a highly deformed and metamorphosed gneiss complex whose age exceeds 3.0 Ga. The history of this terrane has proved difficult to decipher because it was formed and later affected by a complex series of multiple intrusions, metamorphisms, and deformations. There are two areas, however—one in the Minnesota River Valley near the towns of Morton and Granite Falls, and the other near Watersmeet in southern Michigan—that have been extensively studied and whose gneisses have yielded early Archean radiometric ages. In both of these areas the gneisses, which were originally emplaced as granitoids, contain inclusions of metamorphic rocks that probably originated as mafic and ultramafic lava flows. Some of these rocks are older and some younger than the gneisses, the younger ones having been folded into the gneisses by intense deformation. The older supracrustal rocks have proven intractable to radiometric dating, but the gneisses have yielded Rb-Sr isochron and U-Pb discordia ages ranging from 3.48 ± 0.11 Ga to 3.68 ± 0.07 Ga.

Barberton Mountain Land, in northern Swaziland and the southeastern part of the Transvaal Province of South Africa, is another well-studied area of early Archean rocks. It contains the Barberton Greenstone Belt, a thick volcanic and sedimentary sequence that is now highly deformed and intruded by granitoids and gneisses. The lower part of this supracrustal sequence includes mafic and ultramafic lavas with some felsic lava flows and chert deposits. The lavas of these two units have been dated by the Sm-Nd and Pb-Pb isochron methods and by the Ar-Ar age spectrum method, with the resulting ages ranging from 3.53 ± 0.05 Ga to 3.46 ± 0.07 Ga. Rb-Sr and Sm-Nd isochron ages on the oldest granitoid gneisses that intrude and enclose the supracrustal rocks range from 3.42 ± 0.03 Ga to 3.56 ± 0.11 Ga, while a tonalite gneiss has a Pb-Pb age from zircons of 3.64 ± 0.002 Ga.

There are other areas in the world where early Archean rocks occur, although not all have been as extensively studied as those previously discussed. For example, samples of the ancient gneisses of the Imataca Complex of Venezuela have yielded a Pb-Pb isochron age of 3.77 ± 0.02 Ga. The Imataca Complex has been correlated with similar Archean gneisses across the Atlantic in Liberia and Sierra Leone, from which whole-rock Rb-Sr ages of 3.1–3.7 Ga have been measured.

Rb-Sr isochrons ranging from 3.42 ± 0.06 Ga for granitoid that intrudes older supracrustal rocks to 3.52 ± 0.06 Ga for granitoid gneisses have been reported from the Rhodesian craton in Zimbabwe. To the south of the Rhodesian craton there are well-documented early Archean Gneisses near the Sand River. Grey gneisses, and the more felsic gneisses that enclose them, give Rb-Sr isochron ages of 3.73 ± 0.06 Ga and 3.78 ± 0.10 Ga, respectively, while basaltic dikes that intrude the gneisses give a Rb-Sr isochron age of 3.57 ± 0.10 Ga.

Zircon crystals from a gneiss in Liaoning Province, China, some 1000 km northeast of Beijing, have U-Pb ages of 3.80 ± 0.005 Ga and 3.81 ± 0.004 Ga. Ion probe analyses of the cores of zircon crystals in gneiss from Mount Sones, Enderby Land, Antarctica, show that the original granitoid from which the gneiss was formed crystallized at 3.93 ± 0.01 Ga.

Ancient Minerals

How much can be learned from a single mineral grain? Quite a lot, as it turns out. In the northwestern part of western Australia's Yilgarn block (see Figure 5.2), near Mt. Narryer, there are metamorphosed sedimentary rocks containing mineral grains older than any known terrestrial rock. Among these sedimentary rocks is a metamorphosed sandstone composed of mineral grains that were eroded from pre-existing igneous rocks and then redeposited in Archean seas about 3.3–3.5 Ga. In addition to quartz and other minerals, this sandstone contains crystals of zircon that have been analyzed with the ion probe over a period of nearly two decades as researchers from the Australian National University search for older and older crystals. Most of the zircon grains are 3.5–3.7 Ga in age, but four grains exceed 4.0 Ga. In another ion probe study of 140 zircon grains from a conglomerate in the

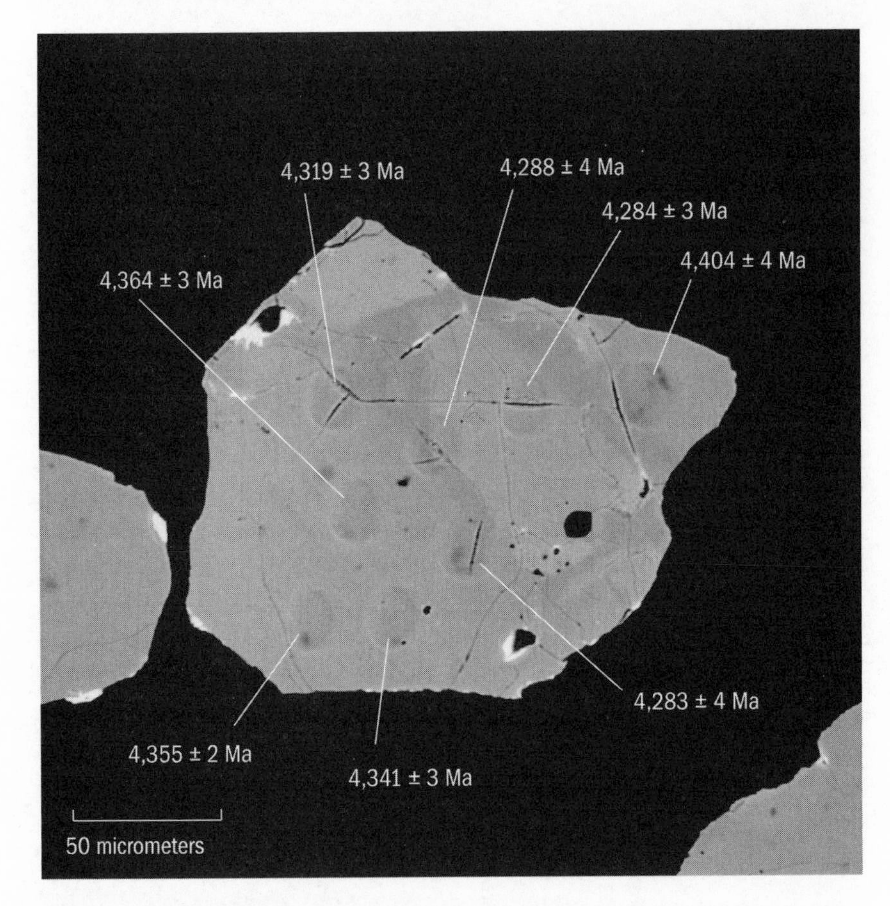

Figure 5.4 The oldest Earth object known. This tiny zircon crystal has a Pb-Pb age of 4.404 Ga. It was found in sedimentary rocks about 3 Ga in age in western Australia. The faint ovals are areas eroded by the ion beam during analysis with the SHRIMP ion probe. Different spots have different ages because of lead loss in those particular areas of the crystal. (Photo courtesy of S.A. Wilde and J.W. Valley. After Wilde and others, *Nature*, vol. 409, pp. 175–178, © 2001, with permission from Macmillan Magazines Ltd.)

Jack Hills, located about 60 km northeast of Mt. Narryer, 17 grains with U-Pb ages exceeding 4.0 Ga were found, including one grain with an age of 4.28 ± 0.06 Ga.

Recently the search for ancient mineral grains in the Jack Hills area of Australia was joined by a team headed by Simon Wilde of the Curtin University of Technology in Perth, and John Valley of the University of Wisconsin. Their search paid off—they found a single zircon grain that is older than any other rock or mineral known on Earth. Ion probe analysis of this grain showed that most of the grain is about 4.3 Ga, but one area of the crystal gives a Pb-Pb age of 4.404 ± 0.008 Ga (Figure 5.4; see also Figure 4.7). The age differences within the grain are thought to be real, with the younger ages representing episodes of Pb loss, perhaps due to one or more thermal events. The older age of 4.404 Ga represents the original crystallization age of the zircon crystal because it is defined by a data point on the concordia curve, indicating that this part of the zircon crystal has lost no lead.

A study of oxygen isotopes in these ancient Australian zircon crystals shows that they were eroded from a granitoid that itself had formed by partial melting of continental crust containing supracrustal sediments. Thus, it appears that Earth had developed continents and oceans within 150 million years or so after it formed. It is truly remarkable how much important information clever scientists can glean from a few very tiny crystals!

Although not rocks, the zircon crystals from Australia represent the oldest materials found so far on Earth. Even though the source rocks of these crystals have not been found, scientists continue to search for them, and there is hope that remnants of the Earth's earliest crust may yet be found.

What Does This All Mean?

There are many locations on Earth where there are rocks with radiometric ages between 3.5 and 4.0 Ga, and in Australia there are crystals whose ages extend as far back in time as 4.4 Ga. None of this evidence provides a precise age for Earth's birth, but it does show conclusively that the planet is more than 4.4 Ga in age. Indeed, it appears that Earth had already developed some continental land masses and had abundant water, probably in the form of oceans, 4.4 billion years ago.

We know from other evidence, to be discussed in later chapters, that Earth's age is between 4.5 and 4.6 Ga, yet the oldest rocks found on Earth are only about 4.0 Ga. What happened to the rocks that represent the first half-billion years of Earth's history? The answer to this question is not really known—there are only speculations and possibilities. One possibility is that during that period of Earth's history not only was the first continental crust forming, but it was also being vigorously recycled and regenerated. So it may be that the earliest crustal rocks were consumed by recycling into the primitive mantle almost as fast as they were generated. A second possibility is destruction by large impacts. The Moon and, by inference, Earth were subjected to intense bombardment by asteroids and large meteorites from the time of their initial formation to about 3.8 Ga. The scars from these impacts can still be seen on the Moon but have been erased from Earth's surface by subduction of the crust back into the mantle, erosion, and the deposition of new rocks. Perhaps the bombardment was sufficiently intense to obliterate the earliest crustal rocks. A third possibility is that the record of Earth's early history exists somewhere but has not been found. The reason for the absence of the most ancient rocks may well be some combination of the above. The discovery of zircon grains 4.0–4.4 Ga old in sedimentary rocks in Australia offers the hope that some of the rocks from Earth's earliest history may have survived and eventually will be discovered.

Whatever the reason for the missing record on Earth may be, much can be learned about the history and age of our planet by examining the evidence from more primitive bodies in the Solar System—particularly the Moon and the meteorites. So next stop, the Moon.

CHAPTER SIX

Moon Rocks: Samples from Our Nearest Neighbor

The trips to the Moon by the Apollo astronauts were surely the greatest feats of engineering and exploration in the history of humankind. In addition to the technical and spiritual benefits, the manned lunar missions also had significant scientific value because they gave scientists, for the first time, the exciting opportunity to study the rocks of another planet. We think of the Moon orbiting around Earth, but objects in space actually orbit around a common center of mass, so the distinction between a planet and a moon is somewhat artificial. Invariably, the smaller of the two is assigned the role of "moon." Earth's Moon is quite large and can fairly be called a planet.

What can be learned about the age of Earth by studying the Moon? The answer lies in the fact that the Moon and Earth, along with the entire Solar System, formed at about the same time. Both the theoretical evidence and the experimental evidence are quite good—there is no evidence to the contrary—so the age of the Moon has a direct bearing on the age of Earth. Good evidence also indicates that the Moon formed from debris generated by the collision between the growing proto-Earth and another object about the size of Mars. Thus, the ages of Earth and the Moon should be about the same.

Hundreds of radiometric age measurements on lunar rocks are on record. The majority fall within the range of 3.2–4.0 Ga and the oldest are about 4.5 Ga. Several lunar rocks have ages that exceed those of the very oldest rocks found on Earth, and very few lunar rocks have formation ages younger than Earth's early Archean. Thus, compared to most Earth rocks, Moon rocks are very old, and they represent events early in the history of the Moon. Some of these events, particularly the heavy bombardment that left deep scars on the Moon's surface, may have affected Earth as well as the Moon, but all earthly record of them appears to have been erased by the vigorous recycling of crust that continually reshapes Earth's surface. In a sense, then, the rock record of the Moon and Earth are complementary, each providing detailed records of somewhat different times in the evolution of the Earth-Moon system.

The Origin of the Moon

Over time, four principal hypotheses for the formation of the Moon have been proposed: fission, capture, formation with Earth as a double planet, and collision. The *fission hypothesis* postulates that the rapid rotation of Earth while in a molten state resulted in the ejection of a large "droplet" that solidified in orbit and became the Moon. Aside from the near identity of the Moon's density and the uncompressed density of Earth's mantle, which may simply be a coincidence, there are few arguments in favor of this hypothesis and considerable evidence, both dynamic and chemical, against it.

The dynamic objections to the fission hypothesis are many. First, there is insufficient angular momentum in the Earth-Moon system, by a factor of three, for the primordial Earth to have formed the large bulge necessary for fission. Second, friction within Earth would dampen any tidal bulge before it could become large enough to throw off part of its mass. Third, the size of a moon produced by fission would be about 20% of the mass of Earth rather than the 1.2% it actually is. Fourth, because of its initial nearness to Earth, tidal forces would have quickly destroyed a fission droplet. Fifth, it is highly improbable that a fission droplet would have had precisely the velocity required to stay in orbit. More likely it would have escaped entirely to orbit the Sun or would have fallen back to Earth. Finally, a Moon created by fission would orbit within a few degrees of Earth's equatorial plane. Instead, the inclination of the Moon's orbit varies 18.5–28.5 degrees.

In addition to the physical difficulties, there are serious chemical problems with the fission hypothesis. Compared to the whole Earth, for example, the Moon is depleted in iron, in many of the siderophile ("iron-loving") elements, such as nickel and molybdenum, and in the volatile elements, such as potassium. In addition it is slightly enriched in uranium and the rare earth elements, which include lanthanum (element number 57) through lutetium (element number 71). The composition of the Moon also does not match that of Earth's mantle. For example, the iron content of the Moon is nearly twice that of Earth's mantle, whereas the concentrations of the siderophile elements are much lower. There are other differences as well, and none of them can be explained adequately, leading scientists to conclude that fission is probably not the way the Moon formed.

The *capture hypothesis* envisions a Moon formed elsewhere and gravitationally captured when it passed near Earth. This has one attractive feature—it neatly solves the problem of the chemical differences between Earth and Moon. If the Moon formed elsewhere, then such differences might easily be accounted for. There are, however, some limits on the region of the Solar System in which the Moon can have formed. Analyses of Earth, the Moon, and meteorites, backed by theoretical considerations, have shown that the ratios of certain light isotopes, particularly those of oxygen, vary systematically from region to region within the Solar System because of light-isotope fractionation and nuclear processes in the condensing solar nebula. The ratios of oxygen isotopes in lunar rocks are identical to those in Earth. This fact appears to restrict the formation of the Moon at least to the same region of the Solar System as Earth. But if this is so, then how can the chemical differences between the Moon and Earth be explained? Thus, the chemistry problem is not, after all, solved by capture unless the Moon originated in a completely different star system.

The dynamic difficulties with the capture hypothesis are also severe. Unless there is some mechanism for slowing it down, an approaching body would not go into orbit around Earth but would either collide with it or be deflected and return to an orbit around the Sun. Returning spacecraft accomplish orbital injection by firing rocket engines to brake their approach. Although not impossible, it is a far more difficult feat to slow a planet just the right amount to ensure orbit. In addition, the orbit of the Moon around Earth is nearly circular, which requires an even more exacting set of circumstances. The capture of a body from outside the Solar System adds the additional requirement that the body escape the gravitational field of its own star and enter the Solar System on a path very nearly in the ecliptic plane—that is, the same plane in which Earth orbits the Sun. These requirements make the probability of capture very small.

In the *double-planet hypothesis*, the Moon and Earth formed together as a planetary pair. This concept satisfies the proximity requirement of the oxygen isotope data in a straightforward way and avoids the severe dynamic problems of capture and fission, but the chemical differences between the two planets remain unexplained, because Earth and Moon would have formed from the same material. A variation of this hypothesis is that the Moon grew from a ring of asteroid-sized objects that formed around Earth shortly after Earth formed. All the variations of the double-planet hypotheses require that elements be fractionated during accretion to account for the difference in composition between the two planets, but there are no known processes that would accomplish such a result.

The *collision hypothesis* calls for material from the proto-Earth, which was then perhaps as small as half its present size, to be injected into orbit by the impact of a large body that was from one to as much as three times the size of Mars. This material would then aggregate as the Moon within a few weeks of the collision. Dynamic and geochemical modeling suggests that such an impact might well result in the formation of a Moon with the right mass, composition, and orbit.

Theoretical considerations indicate that numerous bodies the size of Mars or larger probably impacted the proto-Earth within a period of 20–30 million years during the early stages of Solar System formation. The enormous energy of these large impacts would have melted the outer part of the proto-Earth to form a magma ocean. Collision would also account for the large angle of inclination of Earth's equator from the elliptic plane. There are, however, still some unknown details. What are the implications of a magma ocean for early Earth history, and where are the rocks that would have crystallized as it cooled? How many impacts occurred, and what were the sizes and compositions of the impactors? What relative proportions of proto-Earth material versus impactor material were injected into Earth's orbit? Did the ejecta form a smooth disk orbiting the Earth, or was the disk "clumped"?

Although questions remain, the uncertainties in the details of the collision hypothesis are being clarified rapidly by continuing research. Unlike the other hypotheses for lunar origin, there is as yet no substantial objection to the collision hypothesis and much in its favor, so it is currently considered the best explanation for the origin of the Moon.

As should be obvious from this brief summary, the way in which the Moon originated and how Earth was involved in its birth are imperfectly known. At the heart of this dilemma are the lack of detailed chemical data from other parts of the Solar System and the extreme complexity of the dynamic and chemical processes, also imperfectly understood, that lead to the formation of planets from the debris of star formation. It is important to note, however, that unless the Moon was captured from another star system, a very unlikely scenario, all reasonable hypotheses for the formation of Earth, Moon, and Solar System require them to form at very nearly the same time. Thus, there is much to learn about Earth's age from studies of the ages of lunar rocks.

The Lunar Landscape

Anyone who has looked at the Moon has noticed the distinctive light and dark areas that form the face of the "man in the Moon" (Figure 6.1). The difference in brightness is mostly due to surface roughness. The dark areas, called maria (singular = mare), or seas, are relatively smooth lowland plains and do not reflect sunlight as well as the light areas, which are the more heavily cratered highlands.

The maria account for about 17% of the lunar surface, but they are not distributed uniformly. Nearly all maria are on the Moon's near side, the side that we see. Even before the first Apollo missions provided samples of the maria material, scientists had concluded, from the presence of small volcanic domes and vents, lava flow fronts, lava tubes, and lava flow compression ridges, that the maria were formed by the flooding of low-lying areas, primarily impact basins, with voluminous lava flows. The lack of major volcanoes within or near the maria indicates that most of these flows erupted from fissures. These fissure eruptions were similar to the voluminous basalt lava flows of the Columbia River plateau that flooded the northwestern United States 13–16 million years ago and are beautifully displayed in the walls of the Columbia River gorge.

Except for a few large and spectacular craters, like Aristarchus and Tycho, the maria are the Moon's youngest major features. Where they contact the surrounding highlands, the basalt lava flows of the maria can be seen to fill embayments in the older highland topography (Figure 6.2). Another indication of their relative youth is that the maria are much less heavily cratered than the highlands. Despite the fact that the maria are youthful lunar features, they are still ancient by Earth standards. Radiometric ages of basalt samples from maria are all within the range of 3.0–3.9 Ga. Although samples are available from only a few of the maria, crater density and other geologic considerations suggest that mare volcanism may have continued until about 1 billion years ago but at a much diminished rate.

As their name implies, the lunar highlands are upland areas with eleva-

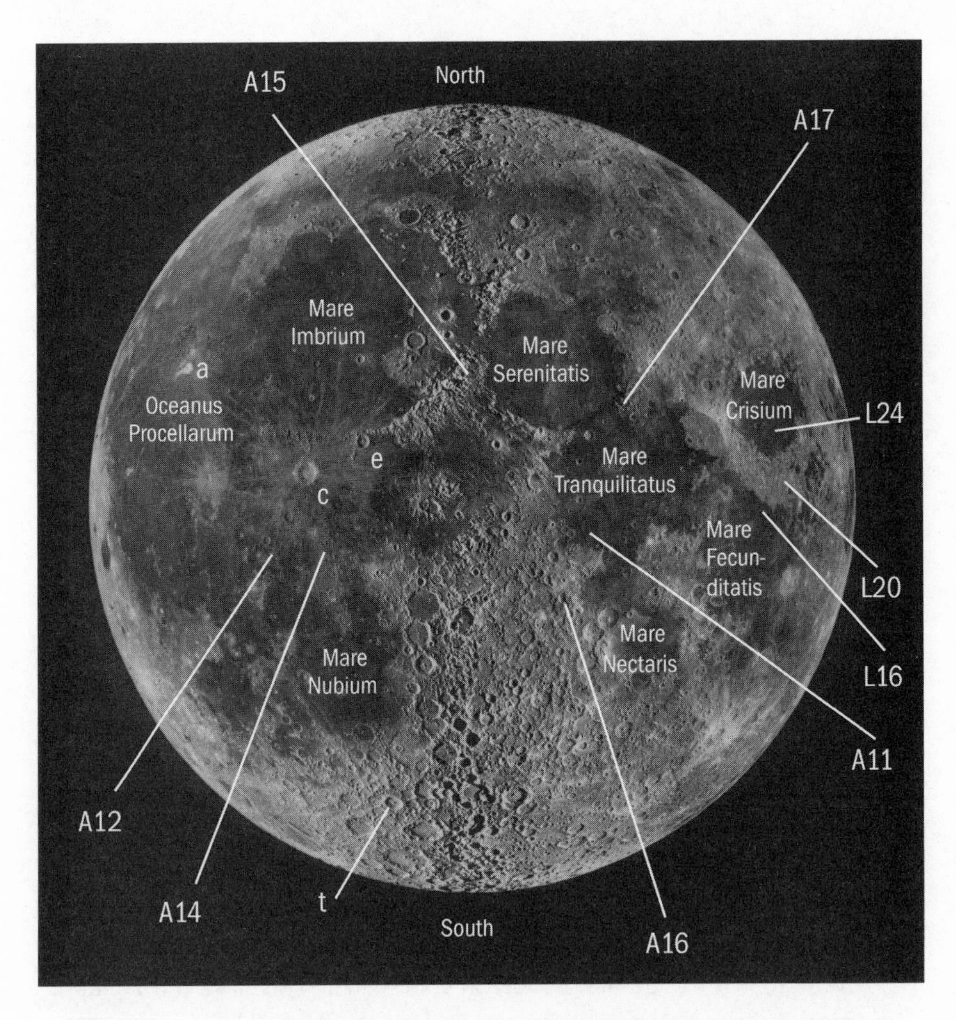

Figure 6.1 The near side of the Moon. This image shows some of the principal named features and the locations of the six Apollo (A) and three Luna (L) landing sites from which samples were returned to Earth. The dark, smooth areas are maria (seas). The brighter, rugged areas are terrae, or highlands. Aristarchus (a), Copernicus (c), and Tycho (t), with their bright rays of ejecta, are three of the Moon's youngest large craters, whereas Erastosthenes (e), whose rays are no longer visible, is older. (Lick Observatory Photograph L-9. Reproduced with permission.)

Figure 6.2 The eastern edge of Mare Serenitatus. This view was photographed by the Apollo 17 mapping camera. The relatively smooth, dark material developed from lava flows that filled the large circular impact basin, which formed about 3.9 billion years ago. There are fewer craters in the mare than in the rugged highlands to the right because the highlands are older. The partly flooded crater Posidonius, about 100 km in diameter, is just north of the older and completely flooded crater Le Monnier. Both craters are visible in Figure 6.1. The Apollo 17 landing site (A17) in the Tarus-Littrow Valley is in the lower right. (NASA photograph AS17-0940.)

tions typically 3–4 km above the maria. The highlands represent the top surface of a lunar crust several tens of kilometers thick. Even before the Apollo landings, scientists knew from the difference in crater density and from stratigraphic relations that the highlands were older than the maria, and radiometric dating has confirmed that conclusion.

Cratering, caused by the impact of rock bodies striking the Moon, has dominated the evolution of the lunar landscape. Lunar impact craters are ubiquitous, and they come in all sizes—from the largest basins, several thousand kilometers in diameter, to tiny micropits, less than one-thousandth of a millimeter in diameter, found on the surfaces of individual mineral grains and rocks.

The basins consist of a central depression surrounded by one or more concentric rings of mountains. The mountain rings distinguish a basin from a crater. There are forty-three basins with a diameter greater than 220 km, distributed more-or-less uniformly over both near and far sides of the Moon. Basins are enormous features that are responsible for most of the Moon's major topography. The largest basin, with a diameter of 2000 km and a relief of some 8 km, is on the far side in the southern hemisphere. The Imbrium Basin, a prominent feature of the northwestern near side, is approximately the size of Texas, and its mountain ranges rise more than 7 km above its floor. Now flooded with basalt lava to form Mare Imbrium, its circular form is still plainly visible (see Figure 6.1).

The lunar basins were caused by the impact of asteroid-sized objects tens of kilometers in diameter. The amount of energy involved in these impacts was enormous. It has been estimated that the excavation of the Imbrium Basin required an energy equivalent to 10 million times the energy released in all the earthquakes on Earth in one year, or enough to energize 10 billion 100-watt light bulbs for 10 million years. This amount of energy corresponds to the collision of an asteroid 100 km in diameter traveling at a velocity of 16 km per second (35,800 miles per hour).

There are no impact features on Earth's surface even remotely comparable in size to the lunar basins, although basins are visible on both Mars and Mercury. Radiometric dating indicates that the lunar basins were formed while the young Moon was collecting the last of the larger bodies in its orbital path. Earth probably was subjected to a similar bombardment at the same time. On Earth, however, any evidence of basins has been destroyed by the constant crustal recycling that is absent on the smaller, now-inactive planets where the basins are still preserved.

The effects of basin formation on the lunar surface are profound. Not only do basins account for most of the Moon's major topography and act as the principal sites for maria, but the frequent and violent collisions pulverized and fractured virtually the entire early lunar crust to depths of about 20 km. As a result, there appear to be no "original" highland igneous rocks, only rock fragments and rocks formed from these fragments by impact fusion. In addition, the material ejected when the basins formed cover large areas of the lunar highlands to depths of hundreds of meters. Indeed, calculations show that such large impacts most likely distributed their ejected material across the entire surface of the Moon. Even the youngest basin ejecta do not occur on the maria, indicating that basin formation preceded the eruption of maria lavas. At one time it was thought that maria lavas were generated by melting due to the basin-forming impacts. The lack of basin ejecta on any of the maria and the great age differences (typically several hundred million years) between basin formation and the maria lavas, however, show that the lava flows came well after the basins formed.

In addition to the basins, the lunar surface is peppered with smaller impact craters whose number increases rapidly with decreasing diameter. There are, for example, about eighty craters with diameters between 100 and 220 km and thousands with diameters between 1 and 100 km. Even though smaller than basins, the larger of the craters are enormous features. Copernicus, for example, with a diameter of 91 km, is 25% larger than the state of Delaware and 3.4 km deep (see Figure 6.1).

The numerous craters with diameters less than 1 km are of both primary and secondary origin. The primary craters were excavated by the impact of bodies striking the Moon, whereas the secondary craters resulted from impacts of lunar rock ejected during the formation of the primary craters. Many of the impacts that formed the smaller craters do not disturb bedrock, but expend their energy in stirring and mixing the material on the lunar surface. They also form new fragmental rocks by fusing together the particles and rock fragments in the lunar soil. These fused fragmental rocks are called breccias, and they are very common in the lunar rock collection.

Microcraters, which range in diameter from about 1 cm to less than a thousandth of a millimeter, are caused by the impact of cosmic dust, primarily comet debris, and of high-velocity ions from the Sun, known as the solar wind. The micrometeorite impacts erode the lunar surface at a rate of about 1 mm per million years, while the solar wind accounts for an additional tenth of a millimeter per million years of wear and tear on the exposed surfaces of lunar rocks.

Because the Moon has been subjected to cratering throughout its history, it is not surprising that the density of primary lunar craters, those with diameters of 1–100 km, is a reasonably reliable indicator of the relative ages of lunar formations. Crater counting is, in fact, one of the most important tools of lunar stratigraphy. Cratering rates, however, have not been uniform for the past 4.5 Ga. Comparisons of the densities of large craters on lunar surfaces of different ages have shown that there was a sharp decrease in impact rates between about 4.0 and 3.5 Ga ago, followed by a further gradual decline in cratering since then. The sharp decline probably represents the final phase in the accretion process that formed the Moon.

Current observations show that cratering on the Moon continues. For example, seismic detectors left on the Moon by the Apollo astronauts record 70–150 earthquakes each year caused by the impact of meteorites with masses between 0.1 and 1000 kg. In addition, fifteen microcraters, caused by the impact of cosmic dust grains, were found on the glass windows of the six Apollo spacecraft that made the journey to the Moon and back.

The heavily cratered lunar landscape may seem exotic, but impact cratering is not unique to the Moon, Mars, and Mercury. Earth, too, is the victim of a similar bombardment. Earth's atmosphere, however, restricts the size and velocity of objects reaching the surface from space, thereby minimizing the frequency and intensity of impacts. Extraterrestrial objects between about a thousand grams and 1 ten-millionth (0.0000001) of a gram burn up in the atmosphere by frictional heating. Particles smaller than 0.0000001 g do not burn up; they are slowed and drift downward through the atmosphere to eventually settle on the surface as dust. It is estimated from the rate at which such particles penetrate satellites that some 880,000 kg (969 tons) of meteoritic dust is deposited on Earth each year, and the dust is common in deep-sea sediments.

Bodies larger than 1000 g occasionally penetrate the atmosphere and impact the Earth's surface as meteorites. More than 150 impact craters ranging in diameter from a few tens of meters to as much as 260 km have been identified on Earth. The Pleistocene-age Meteor Crater (also called Barringer

Crater) in Arizona, the 15 Ma-old Ries Basin in Germany, and the two late Paleozoic craters that form Clearwater Lakes in Quebec, Canada, are well-known examples. A comet or asteroid some 10 km or more in diameter struck the Earth at the end of the Cretaceous period, perhaps causing the extinction of the dinosaurs. The crater it excavated, estimated to be 260 km in diameter, is buried beneath half a mile of younger sediments on the Yucatan Peninsula of Mexico, but it has been mapped by geophysical methods and sampled by drilling. The impact caused enormous ocean waves throughout the Caribbean and injected massive amounts of dust into the atmosphere, causing darkness and the cessation of most plant growth for several years. The dinosaurs may simply have starved.

The incessant bombardment of the lunar surface by rocks and particles from space has resulted in a layer of pulverized rock, mineral fragments, and glass called the lunar *regolith.* About 1.5% of the regolith is meteoritic material and represents the disintegrated remains of impactors. This layer of "soil" covers virtually the entire surface of the Moon (Figure 6.3). The lunar regolith differs substantially from the soils on Earth because the principal processes active on the surfaces of the two planets are very different. On Earth, soils are the result of mechanical, chemical, and biological activity, whereas on the Moon, which lacks an atmosphere, water, and living organisms, soil-forming processes are almost entirely mechanical.

Moon Rocks

One of the most significant scientific benefits of the Apollo program was the return of samples of rock and soil for study by scientists. Six manned missions from the United States and three unmanned missions from the USSR brought back nearly 382 kg (841 pounds) of samples. This priceless material consists of crystalline rocks, breccias, and soil from a variety of geologic environments. Outcrops of undisturbed rock are rare on the Moon, and all the lunar samples were collected from the regolith; none were collected from their exact places of origin, although many probably originated nearby. All were excavated by impacts and many have been moved repeatedly, some for considerable distances. Pieces of highland rocks, for example, have been found on the open maria many tens of kilometers from the nearest highlands.

Moon rocks are not exactly like Earth rocks, and much has been made of the differences. Although these differences are scientifically important, the

Figure 6.3 The lunar regolith. Geologist Harrison H. "Jack" Schmitt, the only scientist to land on the Moon, examines the large boulder at Station 6 near the Apollo 17 landing site in the Tarus-Littrow Valley. The complex breccia boulder rolled downhill from the nearby highlands. The ubiquitous lunar regolith, or soil, composed of pulverized rock, glass, and rock fragments, is the product of repeated meteorite impacts and of radiation from the Sun and space. None of the rocks visible in the photograph are where they originally formed. Note the numerous small impact craters in the regolith. (NASA photograph AS17-140-21497.)

overall similarity of Earth and Moon rocks is even more striking. Contrary to the impression conveyed by many Hollywood films and TV series, there are no totally new or weird types of rocks in the lunar sample collection (Table 6.1). Lunar rocks include both crystalline igneous rocks and impact breccias. Virtually all lunar rock types have terrestrial analogs, although not necessarily in the same abundance. For example, basalt is a common lava

Table 6.1

Simplified Classification of Lunar Highland Rock Types

Rock Name	Relative Abundance	Principal Minerals (%)		
		Plagioclase	Pyroxene	Olivine
Basalts	2	50	35	10
Cumulates				
Anorthosite	3	95	4	0
Anorthositic gabbro	1	70	20	9
Norite (gabbro)	4	40	40	5
Troctolite	5	5	35	60
Dunite	6	2	2	95

NOTE: Plagioclase is a felsic mineral, whereas pyroxene and olivine are mafic minerals. All of these rock types are also found on Earth.

SOURCE: Dalrymple 1991.

type on both Earth and the Moon, and the lunar maria basalts look very much like Earth basalts. Lunar basalts contain the same basic minerals as terrestrial basalts and many contain gas bubbles, or vesicles, just like their Earthly cousins. In fact, geologists who have not studied lunar rocks would be hard-pressed to tell a piece of lunar basalt from a piece of unweathered Hawaiian basalt. Similarly, impact and volcanic breccias are found on both Earth and the Moon. On the Moon, impact breccias are very common and volcanic breccias are rare, whereas the reverse is true on Earth.

The lunar maria are formed of basalt lava flows. Most of these lava flows were emplaced after the early, intense lunar bombardment; as a result, relatively undisturbed samples of maria basalt, with their original igneous textures and mineralogy intact, are common. In contrast, the original crustal layering and lateral distribution of highland rock types has probably been destroyed. All of the larger rocks returned from the highlands are breccias (Figure 6.4). Many of the breccias were formed by multiple impacts, with breccia fragments enclosed within breccia fragments. In the process of breccia formation, some of the rocks were subjected to impact-generated temperatures of as much as 1100° C or more, resulting in incomplete melting followed by recrystallization. Thus, many lunar breccias are really metamorphic rocks formed by multiple events, and many of these rocks are extremely complex. The breccias record a rich history of lunar events, although that history is sometimes difficult to decipher.

Figure 6.4 A lunar breccia. This is rock 15459 from the rim of Spur Crater near the Apollo 15 landing site in the eastern Mare Imbrium. Lunar breccias, composed of a glassy matrix enclosing angular fragments of rock, are formed by impact fusion. The largest visible rock inclusion is about 8 cm across. (NASA photograph AS15-71-44181.)

Samples of the highland rock types occur primarily as inclusions in the breccias and as small fragments in the regolith, although larger samples representing all of the major rock types have been found. Only rarely do highland samples still retain their original igneous textures, because most have been modified by impact processes.

Highland igneous rocks can be grouped into two general categories: basalts and cumulates (see Table 6.1). As their name implies, the basalts formed by the cooling and crystallization of lava. The cumulates, however, represent rocks formed primarily by the concentration of crystals within cooling magma by floating or settling, so their origin is in part due to mechanical processes.

The basalts are primarily of two types that are very similar in composition. The major difference between them is that one type is higher in potassium, the rare-earth elements (lanthanum through lutetium), phosphorus, and such elements as uranium and thorium, and somewhat lower in aluminum than the other type. The former has been given the humorous acronym KREEP (potassium [K], rare earth elements, and phosphorus); the latter is called aluminous basalt, or low-K Fra Mauro basalt.

The cumulates form a range of igneous compositions that grade one into the other, so the simple classification in Table 6.1 is somewhat artificial but convenient. Anorthositic gabbro appears to be the most common highland rock type. It is present at all Apollo and Luna sites, and its composition resembles the average composition of the highlands as measured by orbiting X-ray instruments.

Anorthosite is found mainly as soil and breccia fragments and is sufficiently abundant that it is thought to form a significant percentage of the lunar crust. The existence of anorthosite magma, however, is considered highly improbable because of the extremely high temperatures that would be required to form it. Thus, the anorthosites are thought to be cumulates.

With an increase in pyroxene and olivine and a corresponding decrease in plagioclase feldspar, anorthosites grade into norites, troctolites, and dunites. The textures of these rocks indicate that they are cumulates, similar in many respects to those found in terrestrial gabbro intrusions, like the Stillwater Complex of Montana.

The Geologic History of the Moon

The geologic time scale for the Moon is not nearly as detailed as the one for Earth (Table 6.2). The lunar time scale consists of only five geologic periods, none of which have been further subdivided. The difference in detail between the lunar and terrestrial geologic time scales is due to a number of factors, the most important of which are the difference in the degree of access that geologists have to the rock formations of the two planets, and the complete absence of fossils on the Moon.

Geologists studying Earth can walk at will on its surface, map the distribution and sequence of its rock types, and collect large numbers of samples, including fossils, for detailed examination and analysis—and have been doing so for more than two centuries. In contrast, geologic mapping of the Moon's formations began in earnest in the 1960s and was done using photographs taken through a telescope, later supplemented by orbital photographs and by the few precious samples collected from the nine landing sites. Thousands of geologists have trod the Earth's surface in their effort to decipher its history; only one has ever set foot on the Moon (see Figure 6.3). Nonetheless, the geologic map of the Moon, from which the lunar time scale is derived, is remarkable in its detail considering the handicaps under which it was made.

Table 6.2

Geologic Time Scale for the Moon

Period	Age (billion years)
Copernican	0.0–1.0
Erastosthenian	1.0–3.2
Imbrian	3.2–3.8
Nectarian	3.8–4.0
Pre-Nectarian	4.0–4.6

The relative ages of the various rock units of the Moon were determined using, by and large, the same general principles as those used to study Earth. Foremost among these were the laws of superposition and cross-cutting relationships, which enable an observant geologist to determine the order of emplacement of rock formations at a given locality, though not their ages in years (Figure 6.5).

On Earth, the presence of fossils has enabled geologists to make planet-wide correlations of rock units, resulting in an especially detailed time scale for the past 570 million years. Although fossils are absent, the Moon is not entirely devoid of regional and planet-wide time markers. The widespread deposits of material ejected by basin-forming impacts, in particular, provide a basis for the subdivision of lunar geologic time. Likewise, the flooding of basins by lava flows, forming the maria, provide regional time markers. One of the most useful tools available to the lunar geologist is the density of craters on the various surfaces. Although not linear with time, the density of craters increases with the age of the surface, and crater density is the only quantitative method of determining the relative age of lunar surfaces and their underlying formations on a planet-wide basis.

Copernican rocks include the deposits of the fresh, brightly rayed craters such as Tycho and Aristarchus (see Figures 6.1 and 6.5). The Eratosthenian period includes the deposits of craters like Eratosthenes, whose once-bright rays have been erased by mixing of the regolith and by radiation-induced changes over time. This period also includes one-third of the near-side maria lavas, primarily those in Oceanus Procellarum and Mare Imbrium. The rocks of the Imbrian period consist of the ejecta from the Orientale (on the backside and not visible from Earth) and Imbrium basins. These deposits cover a large part of the lunar surface and provide major marker horizons.

Figure 6.5 Some lunar features of different ages. This photograph was taken by the Apollo 15 mapping camera. Ridges of older highlands peek through the lava flows of the mare Oceanus Procellarium. The flows have flooded the older Prinz Crater. In contrast, the Aristarchus Crater, whose bright "rays" of ejecta form streaks across the mare, is younger than the lava flows. Aristarchus, 43 km in diameter and one of the youngest and brightest of the Moon's impact craters, is also visible in Figure 6.1. The sinuous rills, most of which originate in small craters, are probably lava channels that may have contributed some of the lava that fills the mare. (NASA photograph AS15-2606.)

About two-thirds of the mare lavas are also included within the Imbrian. The Nectarian period includes all deposits that predate the formation of the Imbrium Basin down through and including the deposits of the Nectaris Basin. The deposits of the pre-Nectarian are mostly on the lunar far side.

There is still much uncertainty about the earliest history of the Moon. How, for example, did the Moon acquire a planet-wide crust of anorthosite composition? This could result from the gravitational sinking of the denser iron and magnesium minerals in a magma ocean as it cooled and crystallized, leaving the lighter feldspar crystals to concentrate near the top. But how does a magma ocean form? It is not as fanciful as it might sound at first. What is required is that the Moon formed rapidly. Given that, the gravitational energy released by the impacts of bodies striking the lunar surface during the final stages of its growth would accumulate faster than the heat generated could be radiated into space or conducted into the Moon's interior. The result would be melting of the lunar surface to a depth of several hundred kilometers—a magma ocean from which an anorthosite upper crust would then develop. Partial melting of this anorthosite crust then generated the highlands basalt flows.

Basin formation probably began as soon as the crust had solidified sufficiently to preserve topography, and continued until the Moon had acquired all the large planetesimals in its orbital path, a period of some 600–700 million years. The prevailing view is that basin formation during this period was continuous but declining. This is appealing because it is congruent with the continuous decline in cratering and calls for no unusual events.

An alternative hypothesis, for which there is substantial evidence, is that there was a burst of impacts of large bodies that resulted in what has been called the "terminal lunar cataclysm" about 3.9–3.8 Ga. This would explain the predominance of 3.8–3.9 Ga ages for highland rocks, but it requires that a supply of impacting bodies be available to ravage the Moon some 600 million years after it formed. Such an event might occur either by chance or as the Moon receded from Earth, collecting material from its expanding orbit.

Isotope dating of mare lavas shows that mare volcanism occurred over a period of at least 1 billion years, but morphology, crater density, and the relationship to several young, brightly rayed craters indicate that some mare lavas may be as young as 1 billion years. At the other end of the scale, a fragment of basalt resembling mare lava found in the material ejected from the Imbrium Basin has been radiometrically dated at more than 4.1 Ga. Although mare volcanism and basin formation overlapped in time, the great age differences between basins and their mare fill indicate that mare lavas were not generated by the basin-forming impacts but by much later partial melting of the lunar mantle. Eventually the lunar interior cooled and was no longer able to supply magma, volcanism ceased, and the planet became largely inactive.

For the past billion years or so, the Moon has been quiescent. Its surface has been modified only by the occasional impact of a meteorite and by the relentless assaults of cosmic dust, radiation, and gravity.

The Ages of Moon Rocks

The majority of the ages of lunar rocks are the result of Ar-Ar incremental heating measurements. This is because the Ar-Ar method is somewhat less laborious and much more economical of sample. Therefore, it can be applied to more samples than the Rb-Sr and Sm-Nd isochron techniques.

Some lunar samples have been dated by two or more methods or by two or more laboratories using the same method. The number and type of age measurements tend to be a function of sample size. Large rocks, from which material is plentiful, typically have been dated by multiple measurements using several methods, whereas very small rocks may have been dated by a single measurement using only one technique. Some lunar samples have not been dated; they have been reserved for other types of studies, or have been set aside in their entirety in order to ensure that an adequate and representative supply of lunar rocks is available for future use.

MARE BASALTS

There are more than 100 radiometric age measurements on samples of mare basalt from seven landing sites. The results show that all these samples are ancient by terrestrial standards, even though the flooding of basins by mare lavas was the most recent of the major events in lunar history. The youngest dated mare basalt, from the Apollo 17 collection, is slightly greater than 3 Ga, but there may be younger mare lavas elsewhere.

The oldest dated mare basalts are 3.8–3.9 Ga in age and were collected by the Apollo 11 and Apollo 17 astronauts. There is evidence, however, of even earlier mare volcanism. Several small fragments of basalt found in the

Fra Mauro Formation closely resemble mare basalts in composition and have ages near 3.9 Ga. The oldest mare basalt candidate, a fragment in an Apollo 14 breccia, has a Rb-Sr isochron age of 4.14 ± 0.05 Ga. These older mare-type basalts suggest that the earliest mare volcanism may have preceded the most recent basin-forming events (Imbrium, Orientale) by as much as 300 million years. Although their compositions are similar to mare lavas, it is by no means certain that these basalts originated as the same type of extensive lava "floods" that formed the maria, and they may simply be samples of smaller lava flows.

Most of the mare basalts are pristine rocks, and their radiometric clocks are generally unaffected by later disturbances. Repeated analyses of single samples usually agree to within about 2% or better. When data are carefully selected to include only results of the highest precision, it is possible to resolve age differences of 50 million years or less and to determine the age sequence of basalts of different chemical types.

The concordance of ages measured using different methods and different minerals separated from the same sample is impressive. An example is rock 10072, a large sample of mare basalt collected by the Apollo 11 astronauts. A seven-point Rb-Sr isochron and a six-point Sm-Nd isochron both give ages of 3.57 Ga. Ar-Ar age spectrum dating of this sample has been done on the whole rock and on three different minerals. The two analyses of the mineral plagioclase, which have the most convincing plateaus and provide the best Ar-Ar values for the crystallization age of the sample, agree exactly with the Rb-Sr and Sm-Nd ages. The ages for the whole rock and two other minerals agree with the other results within the analytical uncertainties. Although sample 10072 is one of the best examples because it has been the object of much study, there are many other mare samples whose radiometric ages show equally impressive concordance. Such consistency both among and between methods indicates clearly that the Ar-Ar, Rb-Sr, and Sm-Nd isotope "clocks" in the lunar rocks are not only accurate but are being properly read.

The age results on the mare basalt samples show that mare volcanism was not simultaneous but occurred over a period of at least 1 billion years, probably longer, and may have begun even before the most recent basin-forming impacts 3.8 billion years ago. Within an individual landing site, the age results are distributed over a range of a few hundred million years. The age results also show that the basins were flooded with lavas at different times. The ages for Apollo 12 mare lavas from Oceanus Procellarum, for example, fall between 3.0 and 3.3 Ga, while those from Mare Tranquillitatis (Apollo 11) range in age from 3.45 to 3.9 Ga. The spread in ages from a single site is real and indicates that the filling of individual basins by mare lavas occurred over a long period of time. None of the maria have been sampled at more than a single site or at any substantial depth below the surface, so the age ranges reflected by the samples are minimum ranges for the duration of volcanism in the individual maria.

HIGHLAND ROCKS

It is no surprise that the radiometric ages of highland rocks indicate that, as a group, they are older than the mare lavas. Highland samples, however, are mostly breccias. They tend to be very complex rocks, and the interpretation of their radiometric ages is sometimes difficult. This is because the breccias are both fragmental rocks, composed of bits and pieces of pre-existing rocks, and metamorphic rocks, reconstituted and modified by the heat and fusion of one or more impacts. It is not uncommon for lunar breccias to contain fragments of earlier breccias, and as many as five generations of breccia within breccia have been found in a single sample. Thus, a radiometric age on a breccia sample may represent the time(s) of breccia formation, the time(s) of crystallization of the rock fragments it contains, or something in between.

Figure 6.6 includes 359 radiometric ages for more than 100 lunar highland samples. Some of the ages in the figure are for different rock inclusions from the same breccia, and some are for different rock fragments from the same soil sample. Thus, the number of distinct lunar rocks represented in the figure is considerably greater than 100. A great deal of early lunar history can be deduced from these results, including the ages of many of the basin-forming impacts.

The Apollo 14 samples are mostly of the material ejected by the Imbrium impact. The highland samples in the Apollo 15 collection are also thought to be primarily from the Imbrium Basin. The age of the Imbrium material, and hence of the Imbrium impact that deposited it, must be equal to or younger than the ages of its constituents. The exceptions are those samples that may have been disturbed by later, smaller impacts or that represent material from the formation of the younger Orientale Basin, a thin layer of which may cover the Apollo 14 site. Most of the Apollo 14 and 15 samples have ages between 3.8 Ga and 3.9 Ga, and there is a pronounced mode at

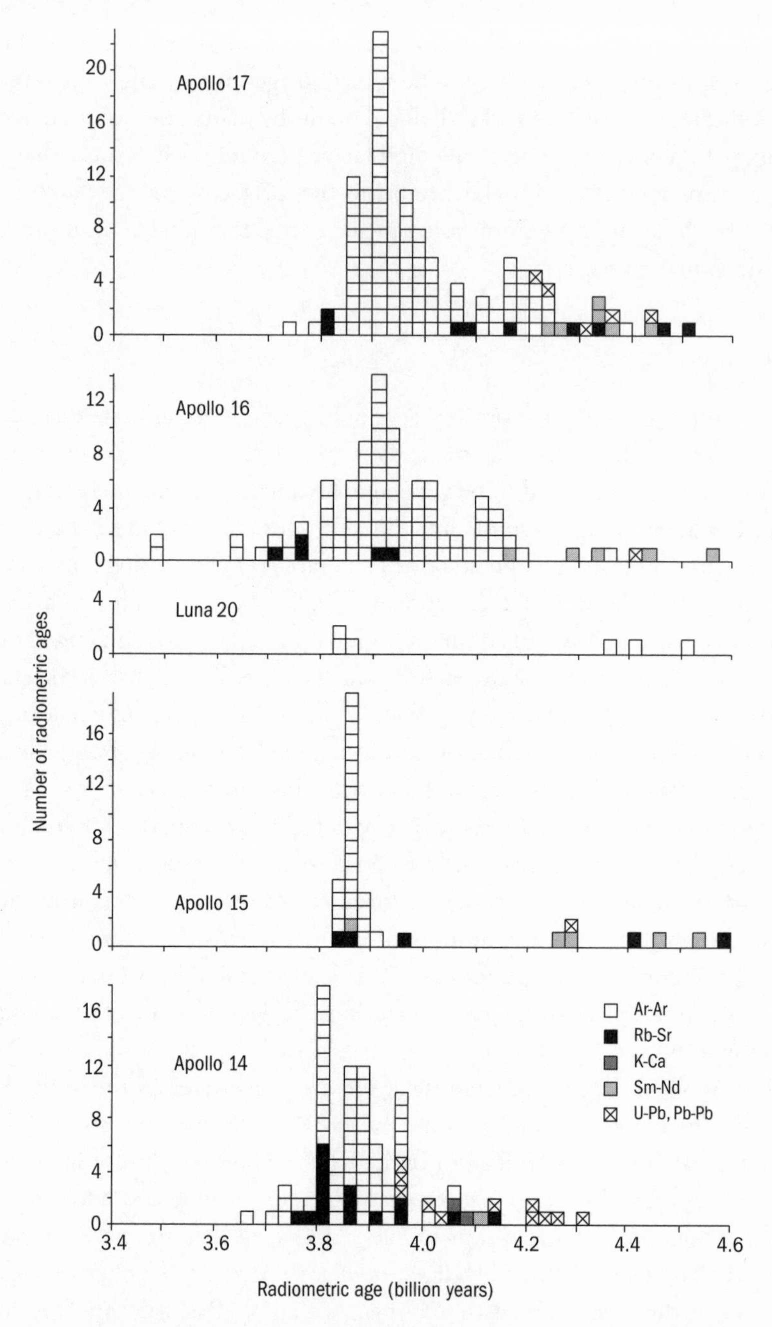

Figure 6.6 Radiometric age measurements on samples from the lunar highlands. The dated rocks include both igneous crustal rocks and rocks formed by basin-forming impacts.

about 3.82 Ga, which provides an approximate older limit for the age of the Imbrium Basin.

Specific samples also yield useful information. Sample 14310, for example, is a crystalline rock, the largest collected on the Apollo 14 mission, that appears to have been formed by the heat of impact. Thirteen Ar-Ar and Rb-Sr isochron ages on this sample have a mean of 3.83 Ga. There is no assurance that this rock was produced by the Imbrium impact—it could have been formed earlier—so its age, too, is an older limit for the Imbrium impact event.

Within the last decade a colleague, Graham Ryder of the Lunar and Planetary Institute, and I have developed methods of determining precise radiometric ages for some of the basin-forming impacts. Such large and highly energetic impacts invariably create a pool of melted rock within the impact crater that may be hundreds of meters thick and many kilometers in diameter. This melted rock then cools and crystallizes, which resets its Ar-Ar clock. So measuring the age of a melt rock also dates the time of the impact that formed it. Until recently, however, Ar-Ar ages on melt rocks have been questionable, because they contain numerous fragments of unmelted and partially melted rock. These older fragments contaminate age measurements and result in radiometric ages that may be too old. Now, however, it is possible to pick out and measure the ages of tiny pieces of pure melt rock. These tiny samples, which typically weigh less than a thousandth of a gram, are progressively heated with a laser beam while using an infrared microscope to measure their temperature. The argon released by the heatings is then analyzed to yield an extremely detailed and reliable Ar-Ar age spectrum despite the small sample size.

The results of the laser Ar-Ar age measurements on more than thirty melt rock samples from the Apollo 15, 16, and 17 sites have yielded some very interesting results. Perhaps the most interesting finding is that there is no evidence for basin-forming impacts prior to 4.0 Ga. This is consistent with the idea of a terminal lunar cataclysm 3.8–3.9 Ga, but inconsistent with the notion that many large objects bombarded the Moon prior to 4.0 Ga. The results on Apollo 15 melt rock samples dated with the laser system show that the Imbrium Basin is probably no older than 3.84 Ga.

There is little or no direct evidence for the age of formation of the Orientale Basin, but based on the probable presence of a thin layer of Orientale ejecta at the Apollo 14 site, the few breccias with post-Imbrium ages may be

Orientale ejecta. If so, then the Orientale impact may have occurred between about 3.75 and 3.81 Ga.

There are very few ages on the Luna 20 material from the Apollonius highlands because of the small amount of sample returned to Earth by the unmanned spacecraft. There is a cluster of Ar-Ar ages at 3.84–3.85 Ga. Geologic mapping indicates that the Luna 20 site is probably dominated by material from the impact that formed Mare Crisium. If the Luna 20 samples do represent material from the formation of the Crisium Basin, then Crisium must be either equal to or less than 3.84 Ga in age, very nearly the same age as Imbrium.

The Apollo 16 site is dominated by highlands samples with ages of about 3.90–3.92 Ga. Hypotheses for the origin of the deposits at the Apollo 16 site suggest that they are composed of material of Nectarian or pre-Nectarian age. Preliminary dating of melt rocks using the Ar-Ar laser system indicates that the Nectaris Basin formed about 3.90 Ga. Crater counts and other geologic observations show that at least ten major basins formed between Nectaris and Imbrium, so a terminal lunar cataclysm lasting less than 100 million years seems likely.

Most of the dated Apollo 17 highlands samples have ages between about 3.85 and 4.0 Ga. Many of these breccia samples are quite complex, and it is thought that most of the ages do not reflect the times of breccia formation and impact but represent a combination of contamination by older material and partial impact resetting. Ar-Ar laser dating of pure melt rock samples, however, shows clearly that the Serenitatis impact is 3.89 Ga in age.

THE OLDEST LUNAR ROCKS

There are only a handful of lunar rocks with radiometric ages greater than 4.2 Ga. All except one, a glassy granitoid (73217), are cumulates. Most were found as inclusions in breccias and all probably endured one or more episodes of post-crystallization impact heating. Because of their complex histories, the interpretation of the data is not straightforward, but each rock has a story to tell, as two examples will illustrate.

Sample 77215 is a piece of brecciated norite that was enclosed within a large breccia boulder sampled by the Apollo 17 astronauts. The norite has been dated by Sm-Nd and Rb-Sr isochron techniques, which give ages of 4.37 Ga and 4.33 Ga, respectively (Table 6.3). All of the Rb-Sr data fall on

Table 6.3

Radiometric Ages of Some of the Oldest Lunar Highland Rocks

Mission	Sample Number	Rock Type	Method	Age (billion years)
Apollo 15	15445	Norite	Sm-Nd	4.46 ± 0.07
	15455	Anorthosite	Rb-Sr	4.42 ± 0.10
			Rb-Sr	4.59 ± 0.13
			Sm-Nd	4.53 ± 0.29
Apollo 16	60025	Anorthosite	Sm-Nd	4.44 ± 0.02
			Pb-Pb	4.42 ± 0.07
	62236	Anorthosite	Sm-Nd	4.36 ± 0.03
			Sm-Nd	4.29 ± 0.06
			Ar-Ar	3.92 ± 0.04
	67016	Anorthosite	Sm-Nd	4.56 ± 0.07
	67435	Plagioclase	Ar-Ar	4.35 ± 0.05
Apollo 17	72417	Dunite	Rb-Sr	4.47 ± 0.10
	73217	Granitoid	U-Pb	4.36 ± 0.02
	76535	Troctolite	Sm-Nd	4.26 ± 0.06
			Sm-Nd	4.33 ± 0.06
			Rb-Sr	4.51 ± 0.07
			Pb-Pb	4.23 ± 0.04
			K-Ar	4.27 ± 0.08
			Ar-Ar	4.19 ± 0.02
			Ar-Ar	4.19 ± 0.02
	77215	Norite	Sm-Nd	4.37 ± 0.07
			Rb-Sr	4.33 ± 0.04
			Ar-Ar	3.92 ± 0.03
			Ar-Ar	3.99 ± 0.03
	78235	Norite	U-Pb	4.43 ± 0.06
	78236	Norite	Sm-Nd	4.34 ± 0.05
			Rb-Sr	4.29 ± 0.02
			Ar-Ar	~ 4.36
Luna 20	L2015	Anorthosite	Ar-Ar	4.40 ± 0.10
	22013	Dark Anorthosite	Ar-Ar	4.36
		Light Anorthosite	Ar-Ar	4.51

NOTES: All ages are based on the isochron (Rb-Sr, Sm-Nd), concordia-discordia (U-Pb), or Ar-Ar age spectrum method. All dated samples were found as inclusions in larger pieces of breccia or as fragments in the lunar soil.

SOURCE: Shih et al. 1993; Alibert et al. 1994; Carlson and Lugmair 1988; Premo et al. 1999; Borg et al. 1997; Norman et al. 1998; Premo and Tatsumoto 1991,1992; compilation in Dalrymple 1991.

a well-defined isochron except for two of the pyroxene measurements. This may be due to disturbance of the Rb-Sr system by the breccia-forming event or by later impact heating. Ar-Ar ages on whole-rock and plagioclase samples from 77215 give ages slightly less than 4.0 Ga, indicating that the norite breccia was affected by later heating sufficient to reset or partially reset the K-Ar system.

The simplest interpretation of the age results for 77215 is that the norite crystallized about 4.35 Ga and was incorporated into the breccia by an impact event that occurred several hundred thousand years later. This interpretation is consistent with Ar-Ar ages of 3.73–3.93 Ga on melt rocks that enclose and cut the norite breccia inclusion. An alternate explanation for the age results is that the Sm-Nd and Rb-Sr ages represent the time the norite was incorporated into the breccia boulder, and that the younger Ar-Ar ages reflect even later impact events. Thus, the crystallization age of the norite is uncertain, but it must be equal to or greater than about 4.35 Ga.

Sample 76535 is a coarse-grained troctolite cumulate whose mineralogy and texture indicate that it formed at depths of 10–30 km in the lunar crust and cooled at a rate of only a few tens of degrees Celsius per million years. The rock shows no evidence of alteration by impacts, although logic dictates that it must have been excavated from depth by a major impact.

The chronology of 76535 has been investigated by Sm-Nd, Pb-Pb, Rb-Sr, conventional K-Ar, and Ar-Ar techniques with some perplexing results (see Table 6.3). The Sm-Nd and Pb-Pb isochron ages, the four Ar-Ar ages, and a conventional K-Ar age are all between 4.2 and 4.3 Ga. The Rb-Sr isochron age, however, is 4.51 ± 0.07 Ga. The cause of this discrepancy is not fully understood, but one explanation is that the Rb-Sr system in the mineral olivine actually resides in microscopic inclusions of other minerals within the olivine grains. According to this hypothesis, the olivine acts as an impenetrable barrier to rubidium and strontium migration even during intense metamorphism or reheating. If this is so, then the Rb-Sr age probably represents the crystallization age of this rock and the younger ages are due to elevated temperatures within the lunar crust between 4.5 and 4.2 Ga, after which the rock was excavated from depth by impact. The Rb-Sr age aside, the Sm-Nd, Pb-Pb, K-Ar, and Ar-Ar results leave no doubt that 76535 is at least 4.2 Ga in age and has been unaffected by impact heating since that time.

There are other highland rocks with radiometric ages exceeding 4.0 Ga, but in every case the interpretation of the results involves some ambiguity.

The best candidates for the oldest lunar rocks (see Table 6.3) include norite 15445 (4.46 Ga), anorthosite 15455 (4.42 Ga), dunite 72417 (4.47 Ga), the two Luna 20 anorthosites L2015 (4.40 Ga) and 22013 (4.51 Ga), and possibly troctolite 76535 (4.51 Ga). If these ages represent crystallization and cooling within the lunar crust, then differentiation of the primeval lunar magma ocean and the formation of the cumulates must have begun about 4.5 Ga or before. Even if these ages represent the times of impact or metamorphism, they still provide minimum ages for the formation of the Moon. Thus, even the most conservative interpretation of the age data on highland rocks leads to the conclusion that the Moon's age must equal or exceed 4.5 Ga.

CHAPTER SEVEN

Meteorites: Ancient Wanderers of the Solar System

We are all fascinated by those thin white streaks of light that occasionally move across the clear night sky at seemingly impossible speeds, then vanish as suddenly as they appeared. Although most people call them "falling stars," they are *meteors,* caused by frictional heating and ionization of air molecules as a tiny bit of cosmic debris, a *meteoroid,* penetrates Earth's atmosphere at very high velocity. Nearly all meteoroids disintegrate in their brief flight through the upper atmosphere and reach Earth's surface only as meteoritic dust. A few of the larger meteoroids, however, survive their fiery passage and strike the Earth, becoming *meteorites.* These occasional visitors from space are more than curiosities; they are valuable samples of objects created when the planets formed—bits of debris left over from the process. Many contain droplets of matter formed by condensation directly from the Solar Nebula and therefore represent some of the most primitive solid material in the Solar System. Others are fragments of large bodies that were chemically differentiated into crust, mantle, and core, and that were obliterated by one or more violent collisions with similar objects. Thus, meteorites record much about the age and early history of the Solar System.

The notion that stones occasionally fall from the sky is now generally accepted and the scientific evidence for the phenomenon is conclusive, but this was not always so. Until the early nineteenth century, the origin of meteorites was uncertain and confused by myth and mysticism. On occasion, stones from the sky have even been venerated as objects of religious significance. Several have been found carefully interred at Native American burial grounds, and the Black Stone built into the Kaaba in Mecca, which was enshrined before Muhammad conquered the city in A.D. 624, reportedly fell from the sky and is probably a meteorite. The fall of meteorites is no longer mysterious, and it is now clear that the collision of cosmic debris with Earth is an ongoing, daily process. Recent estimates of the mass of meteoritic influx based on satellite penetration studies, meteor surveys, and the meteoritic content of deep-sea sediments indicates that Earth receives about 20 million kg of cosmic debris each year, mostly in the form of meteoritic dust. Although this may seem like a large mass of material, uniformly distributed over Earth's surface and left undisturbed for a million years, it would amount to a layer only about 0.02 mm thick.

As might be expected, the number of meteoroids entering Earth's atmosphere each year is a function of their size, with the larger bodies being considerably less numerous than the smaller ones. Objects with masses of 1 trillion kg or greater enter Earth's upper atmosphere only about once every million years. Such an object would have a diameter of 1 km or so, and if it struck land would create a crater more than 50 km in diameter. In contrast, more than 100,000 meteoroids with masses of 1 kg or more enter the upper atmosphere each year. The rate for particles weighing a billionth of a gram or more is about 100,000 trillion per year.

Of all of the meteoritic material that reaches Earth's surface each year, less than 1% arrives in pieces large enough to be called rocks. The fate of a meteoroid entering the atmosphere is primarily a function of its mass and velocity. Objects with masses between about 1 kg and 1 ten-millionth (0.0000001) of a gram disintegrate in the atmosphere, whereas smaller particles are merely slowed and reach Earth's surface carried by air currents. Only meteoroids with masses above about 1 kg completely penetrate the atmosphere and impact Earth's surface. Not surprisingly, composition plays a role in meteoroid survival. The iron and harder stone meteoroids are more durable than the fragile stones.

On average, only about one object large enough to form a substantial crater reaches Earth's surface each year, and large meteorites capable of creating giant explosions strike Earth only once or twice per century. Meteorites that produce geologically durable craters are even rarer. About 160 structures, ranging in age from Cenozoic to Precambrian, are known or suspected to be of meteoritic origin. One of the best known of these is Meteor (or Barringer) Crater near Flagstaff, Arizona, which was formed by the impact of the Canyon Diablo iron meteorite about 50,000 years ago (Figure 7.1, top). An

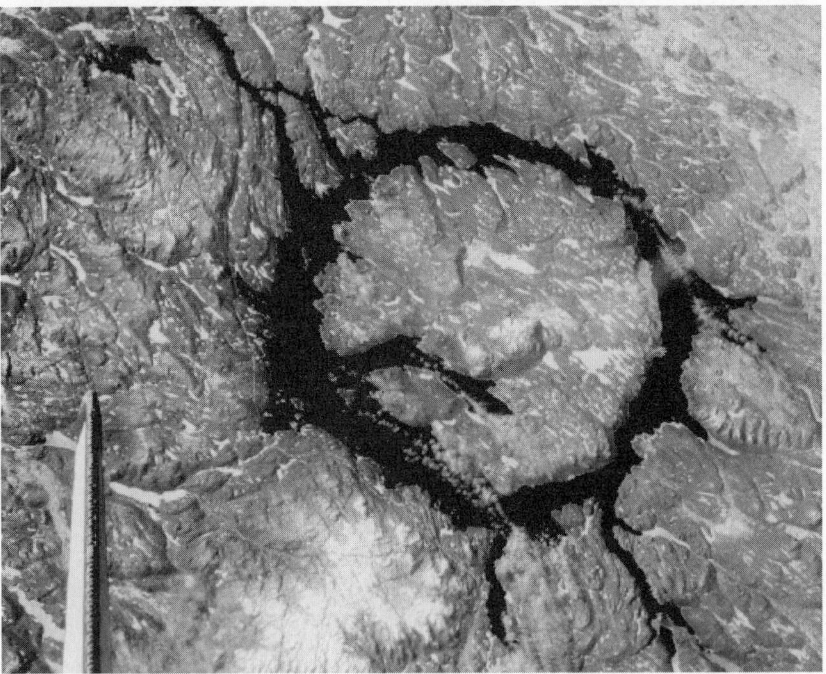

older example is the Manicouagan impact structure in Canada (Figure 7.1, bottom). This interesting remnant of a crater is about 210 million years old and now contains a circular water reservoir.

Meteorites are traditionally named for a geographic feature near the spot where they are found. Hence, the Allende meteorite, which fell in 1969, was found near the little town of Pueblito de Allende in Chihuahua, Mexico. The first pieces of the Canyon Diablo iron meteorite, which excavated Meteor Crater, were recovered near Canyon Diablo, Arizona. Because so many meteorites are found in Antarctica, they are both named and sequentially numbered, with the name being abbreviated. Thus, ALH-761 was found near the Allan Hills and Y-75011 near the Yamato Mountains.

Meteorites are highly prized by both scientists and collectors, and finding one is a rare occasion. Nearly 19,000 meteorites have been collected, and more than 16,000 of these have been found since 1969, the year it was discovered that meteorites are preserved and easily seen on the surface of the Antarctic ice fields. Less than 1,000 meteorites were observed to fall; the others are "finds." Many of the latter are weathered from long exposure to Earth's atmosphere and are therefore not as desirable for scientific study as "falls," which are fresher and provide a more representative sample of the proportions of the various meteorite types that impact Earth.

Figure 7.1 (Opposite) Impact craters. *(Top)* The Meteor (or Barringer) Crater, Arizona, is 1.2 km in diameter and 170 m deep. It was made by the impact, some 50,000 years ago, of the iron meteorite Canyon Diablo. More than 27,000 kg (59,000 pounds) of meteorite fragments have been recovered from the surrounding desert. The crater is young enough to have retained many of its original features. (Photo by D. Roddy, U.S. Geological Survey.) *(Bottom)* The Manicouagan impact structure in Quebec Province, Canada, shown in this 1983 space shuttle photograph, is about 210 million years old. With a diameter of 70 km, it is one of the largest impact craters still preserved on Earth. A ring-shaped lake, the Manicouagan Reservoir, now fills a depression where rock fractured and weakened by the impact has been preferentially eroded away over time. The original crater rim is now gone, but by comparison with other impact structures, it is estimated to have been about 100 km in diameter. The tail of the space shuttle is visible at lower left. (NASA Space Shuttle photograph ST009-48-3139.)

Table 7.1

A Simplified Classification of Meteorites and Their Abundance as Observed Falls

Name	Approximate Percentage of Observed Falls		
Stones	95		
Chondrites		86	
Carbonaceous			5
Ordinary			80
Enstatite			1

Achondrites		9	
Basaltic			6
Others (cumulates and primitive)			3
Stony irons	1		
Irons	4		

NOTES: The dashed line separates the undifferentiated meteorites (above) from the differentiated meteorites (below). The undifferentiated meteorites represent material little changed since formation of their parent bodies from the Solar Nebula. The differentiated meteorites are fragments of rocks that formed by partial melting, crystal accumulation, and differentiation in larger parent bodies.

SOURCE: Dalrymple 1991.

Types of Meteorites

Meteorites are generally classified by the degree to which they are differentiated, or chemically evolved, and by their relative proportions of silicate and metallic minerals (Table 7.1). Stony meteorites, or stones, are composed primarily of silicate minerals and, in a general way, resemble rocks found on the surfaces of Earth, Moon, Mars, Venus, and, presumably, Mercury. Iron meteorites, or irons, are metallic and are primarily composed of an alloy of nickel and iron. Rocks resembling irons are essentially unknown on the surfaces of the major planets, but the core of Earth probably consists of material very much like iron meteorites. In between the stones and irons are the stony irons that are, as their name implies, mixtures of both silicate minerals and metals. Even though this threefold classification is useful, there are intermediate varieties. Thus, many stones contain metal and many irons contain small amounts of silicates.

Based on the proportions of observed meteorite falls, stone meteorites are much more abundant than iron and stony-iron meteorites combined by about 20 to 1. This indicates that stones are much more abundant than irons in Earth's region of the Solar System, but the true ratio may be even higher because stones tend to be more fragile than irons and more of them probably disintegrate in Earth's atmosphere. Of meteorite finds, the combined frequency of irons and stony irons is much greater than that of stones because stones closely resemble ordinary rocks, whereas irons appear unusual and are more apt to be recognized.

CHONDRITES

Stone meteorites are divided into two main types: chondrites, which constitute about seven-eighths of all stone falls, and achondrites. *Chondrites* are so-named because most contain *chondrules*, which are small, rounded mineral clusters that range in diameter from a few tenths of a millimeter to several millimeters (Figure 7.2, top). Chondrules are composed primarily of high-temperature silicate minerals, mostly olivine and pyroxene, with only minor amounts, if any, of metals. Some chondrules are spherical and either glassy or very finely crystalline, indicating that they formed by the rapid quenching of small liquid droplets, whereas others are irregular and coarsely crystalline, indicating slower cooling.

The precise origin of chondrules is not entirely clear and probably not limited to a single mechanism. Those that clearly formed from liquid droplets may have condensed directly from the Solar Nebula, the most commonly accepted hypothesis, or were created by the fusion of silicate dust before accretion of the meteorite parent bodies. Alternatively, they may have resulted from impact melting on meter-sized bodies of rock. Many of the coarsely crystalline chondrules with somewhat irregular shapes may be abraded fragments of once-larger rocks. Whatever their precise origin, it is generally agreed that chondrules represent some of the most primitive solid material in the Solar System.

Chondrules may constitute anywhere from zero to as much as 70% of a chondrite's mass, in a fine-grained matrix of mineral grains and fragments of broken chondrules. Aside from the presence of chondrules, chondrites are distinctive in their chemical composition, which, except for highly volatile elements like hydrogen and helium, closely resembles the composition of the Sun and, presumably, of the Solar Nebula. The primitive, solar compo-

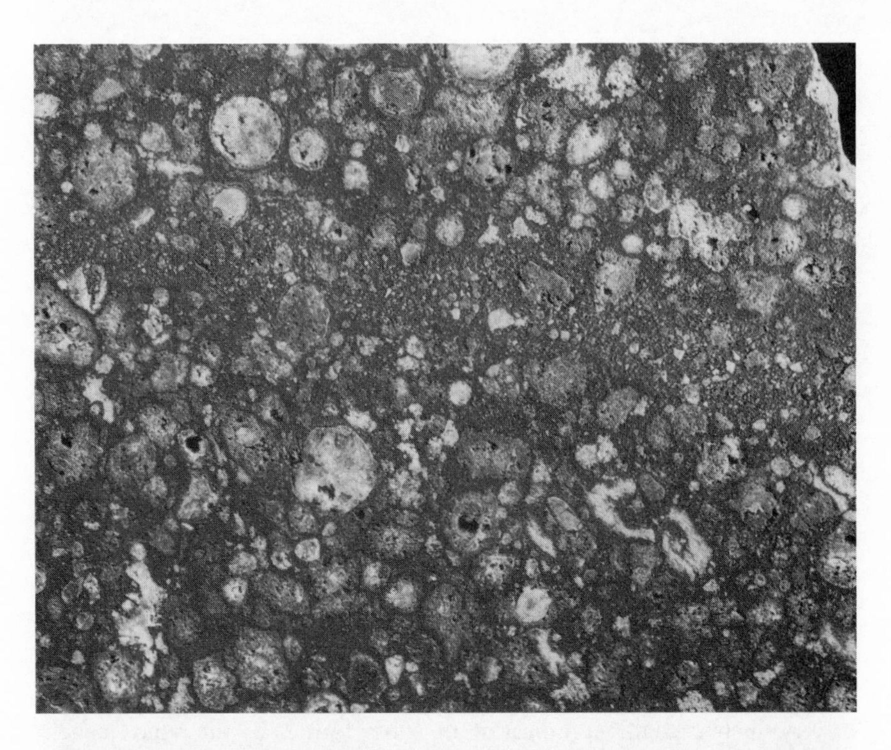

sition of chondrites sets them apart from Earth rocks, Moon rocks, and the other types of meteorites, all of which have been chemically differentiated within the interiors of planets and asteroid-sized bodies.

Chondrites are divided into three classes primarily on the basis of their chemistry. *Carbonaceous chondrites* are both chemically and physically the most primitive of the three (see Figure 7.2, top). Of all of the meteorites, carbonaceous chondrites have compositions that most nearly resemble the Sun. They are relatively high in volatile substances and may consist of more than 20% by weight of water bound within hydrous minerals. They also contain no metallic iron or nickel. In addition, they contain as much as 5% carbon, primarily as graphite, and as much as 1% of organic molecules, including amino acids and long-chain hydrocarbons. Some of the amino acids are the same ones that are associated with protein synthesis in living systems, leading some scientists to speculate that meteorites may have seeded life on Earth. Their high volatile and organic content, their low densities, and their low degree of metamorphism indicate that carbonaceous chondrites have never been subjected to temperatures appreciably above 200° C or to substantial pressures since they formed.

The matrix and the chondrules of the carbonaceous chondrites appear to have formed separately from the nebular gas and dust. Matrix and chondrules then collected to form the chondrite parent bodies. The amount of post-formation metamorphism of these meteorites has been small, involving only mild heat, some alteration by water, brecciation, and slight shock effects in some individual meteorites. All carbonaceous chondrites appear to

Figure 7.2 (Opposite) Two types of meteorites. *(Top)* The Allende meteorite is a carbonaceous chondrite, with spherical chondrules embedded in a fine-grained matrix of mineral grains. Carbonaceous chondrites are the most primitive meteorites, fragments of undifferentiated asteroids that have changed little since they formed. The largest chondrules in the photograph are about 2 mm in diameter. (Smithsonian Institution photograph M-1752H.) *(Bottom)* A cut-and-etched face of the Edmonton, Kentucky, iron meteorite shows the distinctive Widmanstätten pattern, caused by the intergrowth of two different forms of nickel-iron alloy during slow cooling. Iron meteorites are fragments of the metallic cores of asteroids large enough to have differentiated into core, mantle, and crust. The cut face is about 12 cm across. (Smithsonian Institution photograph M-375.)

have had similarly simple post-formation histories at very shallow depths in their parent bodies.

Ordinary chondrites are the most numerous of the observed falls. Although similar in composition to carbonaceous chondrites, they have been subjected to mild metamorphism. The presence of chondrules shows that ordinary chondrites have not been melted since they formed, but their texture and mineralogy indicate that they resided for a while in a dry environment at low pressures and at temperatures of less than 400° C. Such conditions would be found at depths of about 2–5 km beneath the surface of their parent bodies. In addition to differences in volatile content, the ordinary chondrites also differ from carbonaceous chondrites in having less magnesium, calcium, titanium, and aluminum, and more silicon. Although some of the chondrules in ordinary chondrites are crystallized droplets, most are fragments that were probably broken up by impact processes during accretion of the parent bodies.

Ordinary chondrites are subdivided into three separate and distinct compositional groups on the basis of their content of iron and nickel: the H group is high, the L is low, and the LL is very low. More than 20% of all three groups are impact breccias, similar to the lunar breccias, but fragments of one type do not occur in another. Their distinct chemistry and lack of fragment sharing suggests that the three groups originated in separate parent bodies.

The history of ordinary chondrites is more complex than that of carbonaceous chondrites. The nature of their chondrules and the substantial percentage of breccias suggest that impact and shock metamorphism played a more important role in their formation. In addition, post-formation thermal and impact metamorphism clearly affected the ordinary chondrites, whereas such effects on carbonaceous chondrites were minimal. Still, ordinary chondrites are not highly evolved rocks, and their features record relatively early events in the formation of the Solar System.

Enstatite chondrites are the least abundant and the least primitive of the three classes of chondrites. They are mostly impact breccias composed primarily of the mineral enstatite, a type of pyroxene. Their texture and mineralogy indicate metamorphism at temperatures between about 600° C and 870° C. They are chemically similar to some of the achondrites and may have come from the same parent body.

ACHONDRITES

Achondrites differ from chondrites in several ways. They lack chondrules, have compositions that are distinctly nonsolar, and have igneous textures. They formed by the crystallization and cooling of rock melts. Of all the types of meteorites, achondrites most closely resemble terrestrial and lunar rocks. Like their terrestrial and lunar counterparts, the achondrites are the result of differentiation in one or more episodes of partial melting and crystal fractionation within the interiors of their parent bodies. In addition, they were brecciated by impacts and separated from their parent bodies before their collision with Earth.

Achondrites range in composition from rocks composed almost entirely of olivine or pyroxene and resembling terrestrial and lunar igneous cumulates, to rocks very much like terrestrial and lunar basalt lavas. The latter group, the basaltic achondrites, are very interesting rocks, in part because they confirm that basaltic volcanism is common in the inner Solar System. Basaltic achondrites include basalts that crystallized as lava flows (many have vesicles, or gas bubbles) and as shallow intrusive rocks. Their mineralogy, texture, and composition closely resemble terrestrial and lunar basalts. Most of these rocks have been brecciated, and some have been subjected to mild thermal and shock metamorphism. Like the lunar breccias, they were formed on the surface of their parent body by impacts. They are composed of fragments of several types of basaltic achondrites, occasionally with some chondritic fragments.

IRONS AND STONY IRONS

The stony iron meteorites consist primarily of olivine or pyroxene mixed with a nickel-iron alloy. They formed within the interiors of asteroid-sized bodies, perhaps near the core-mantle boundary.

Iron meteorites are composed primarily of nickel-iron alloy (Figure 7.2, bottom). They typically have a distinctive form of crystal intergrowths called Widmanstätten pattern, named after Count Alois von Widmanstätten of Vienna, a codiscoverer of the phenomenon. This unique and beautiful crystal pattern is a result of slow cooling, which would occur in the interiors of large bodies. Some of the iron meteorites no doubt are fragments of the cores of asteroid-sized bodies, whereas others may have originated as metallic segregations at shallower depths.

Where Do They Come From?

Although meteorites were once thought to be the remains of a shattered planet, their chemistry, mineralogy, and textures show that they come from perhaps as many as seventy or eighty separate parent bodies with diameters ranging from about 200 to 600 km. Where are these bodies and how do fragments of them end up on Earth?

There is little doubt that meteorites come from the Solar System. The high frequency of meteors and of meteorite falls requires a voluminous and nearby source. Moreover, synchronized and timed photographs of meteors in flight show that these objects orbited the Sun counterclockwise, like the planets, at low inclinations to the plane of the ecliptic, and typically with the farthest point of their orbit in or beyond the asteroid belt.

Of the three possible sources of meteorites—comets, planets, and asteroids—the evidence indicates that asteroids are by far the largest suppliers. There are many tens of thousands of asteroids in the Solar System with an aggregate mass about 4% of the mass of the Moon. Of these, several thousand have been observed and catalogued. Asteroids range in size from minor planets like Ceres, with a diameter of 1000 km and half the mass of all asteroids put together, to numerous smaller bodies. Estimates of the size distribution of the asteroids vary, but there are at least a dozen with diameters exceeding 250 km, hundreds with diameters greater than 100 km, 10,000 or so whose diameters exceed 10 km, and perhaps as many as 100 trillion larger than 1 meter.

Asteroids occur throughout the Solar System, but the majority are concentrated in the asteroid belt, a diffuse band of rocky material and minor planets between the orbits of Mars and Jupiter. These belt asteroids are debris left over from the process of planet formation and were never part of a single, large planet. The bits and pieces of the asteroid belt could not aggregate into a planet because of their proximity to Jupiter, whose powerful gravity field would have torn apart any body of significant size. Many asteroids appear to be irregularly shaped, a fact first determined from changes in the brightness of reflected sunlight as these bodies tumble in space and only recently verified by spacecraft photographs (Figure 7.3).

Figure 7.3 Asteroids. This is asteroid 243 IDA and its tiny moon, photographed by the Galileo spacecraft on August 28, 1993, from a distance of 10,870 km, while on its way to Jupiter. IDA is about 56 km long and its moon about 1.5 km across. (NASA photograph P-43731.)

Although a large majority of the asteroids occur within the asteroid belt, there are significant numbers elsewhere in the Solar System. The Trojan asteroids, with a probable population of several thousand, orbit the Sun in Jupiter's orbit about 60 degrees ahead of and behind Jupiter. Asteroids also occur in orbits between the asteroid belt and Jupiter, and some, the Hidalgos, occupy orbits beyond Jupiter.

Some of the most interesting are the planet-crossing asteroids, whose orbits intersect those of the inner planets. An estimated 10,000 asteroids travel paths that cross the orbit of Mars and another 500, the Mars grazers, pass very near Mars's orbit. Dozens of these Mars crossers and Mars grazers have actually been observed with telescopes.

Of considerable interest are the Apollo asteroids, which cross Earth's orbit.

Approximately 30 Apollo asteroids with diameters as great as 9 km and about 1000 with diameters greater than 1 km have been observed and catalogued. Because the Apollo asteroids cross Earth's orbit, it is only a matter of time before one of them collides with Earth. Calculations show that, on average, three such bodies with diameters of 1 km or more strike Earth every million years, and that a large asteroid several kilometers in diameter strikes our planet every 40 million years or so. For asteroids 0.5 km or more in diameter and with sufficient energy to produce a crater at least 10 km in diameter, the impact frequency is estimated to be about one every 100,000 years.

Even though asteroids have yet to be sampled by man or machine, a great deal is known about their probable compositions from the technique called *reflectance spectrophotometry.* This method is based on the phenomenon that the amount of sunlight reflected at various wavelengths (colors) from the surface of a body depends on the minerals on the body's surface. The surface compositions of more than 500 asteroids have been determined using reflectance spectrophotometers attached to telescopes. Some three dozen distinct types of spectra have been identified among the asteroids, and most of these closely match the spectra of various meteorites measured in the laboratory.

An interesting result of these compositional studies is that spectra matching the ordinary chondrites, the most common by far of meteorites, are rare among the asteroids—only a few have been identified. In contrast, asteroids resembling carbonaceous chondrites in composition appear to be very common, even though such meteorites are rare on Earth. Why is this so? First, carbonaceous chondrites are very fragile and most may not survive passage through the atmosphere, so despite their abundance in space few may reach Earth intact. Second, the frequency of meteorite types probably depends mostly on the composition of only a few asteroids rather than the entire asteroid population. Interestingly, two of the three asteroids with spectra that resemble ordinary chondrites are Apollo asteroids, whereas few Apollo objects have the spectra of carbonaceous chondrites.

Thus, spectral studies indicate that meteorites have the same mineral compositions as the asteroids, although they represent a statistically biased sample of compositional types. But how do they get here? The orbits and velocities of individual asteroids differ greatly, and because of gravitational interactions with each other and with nearby planets, especially massive Jupiter, collisions between asteroids are inevitable (Figure 7.4). The result-

Figure 7.4 Asteroid collision. This photograph of a painting by W.K. Hartmann shows the collision of two asteroids. The larger one is heavily fractured while the small one disintegrates, ejecting fragments into different orbits. Some of these fragments might be thrown into Earth-crossing orbits by the gravity of Jupiter and strike Earth as meteorites. (From *Moons and Planets*, Second Edition, by W.K. Hartmann, © 1983 by Wadsworth, Inc. Reprinted by permission of publisher and author.)

ing fragments are ejected into new orbits that differ from those of their parent bodies. These initial new orbits are not Earth-crossing, but in time the strong gravitational field of Jupiter gradually shifts some of the fragments into Earth-crossing orbits, where they eventually can (and do) collide with Earth. Thus, the compositions of the contemporary meteorites may be largely governed by the compositions of the last few asteroids to collide and fragment, rather than by the asteroid population as a whole. This means that the fragments themselves are relatively young, even though the bodies from which they come are very old. These conclusions are confirmed by cosmic ray exposure ages, which measure the residence times of individual meteorites in space and range from a few million years to a few hundred million years. Thus, ages of meteorites measured by radiometric dating are really the ages of their parent bodies.

Evidence indicates that chondrites are chips from relatively primitive and small parent objects with diameters of only 100–200 km, whereas the differentiated meteorites probably come from parent bodies as much as 500–600 km in diameter. These larger bodies were heated early in their history by the energy released from the gravity of accretion, the decay of short-lived radioactive nuclides, and perhaps by a strong solar wind. This heating caused melting and differentiation into crust (basaltic achondrites), mantle (cumulate achondrites and stony irons), and metallic core (irons). On some asteroids, magma flowed onto the surface as basalt lava and some of this surface material was converted into breccia by impacts. Successive collisions among asteroids have now exposed them to various depths. Thus, the asteroid Vesta, with a surface composition like basaltic achondrites, is probably intact and its surface covered with basaltic lava flows and breccias. In contrast, the asteroid Athamantis, with a surface composition resembling stony iron meteorites, is probably the residual core of a once-larger body.

Although most meteorites sample asteroids, there is good evidence that a few came from the Moon and Mars. This may seem fantastic, but theoretical studies have shown that it is possible to eject relatively unshocked pieces of rock from planetary surfaces by impacts of asteroid-sized objects, and that some of this material can be injected into Earth-crossing orbits. Asteroids and other bodies that orbit the Sun have quite high velocities. Meteorites typically travel at speeds of 20–40 km per second (22,300–89,500 mph). Earth orbits the Sun at a velocity of 30 km per second (67,100 mph). The velocities of asteroids are similar, so there is easily enough energy to eject rocks from the Moon and Mars by a collision. Mineralogical, chemical, and isotope studies confirm a lunar origin for more than a dozen meteorites collected in Antarctica. Their compositions suggest that they came from several different lunar sites not sampled by the Apollo and Luna missions.

A handful of meteorites appear to be samples of Mars. These rocks are cumulate achondrites that were formed by volcanic activity. They contain gas that is identical in composition to the Martian atmosphere, which was measured by instruments aboard spacecraft that landed on Mars's surface. The ejection of stones from the surface of Mars could result from the impact of an asteroid 10–20 km in diameter traveling at a velocity of at least 40 km per second (89,500 mph), conditions that are within the known size and velocity ranges of asteroids.

Another potential source of meteorites is comets, but for several reasons it is improbable that comets provide any of the meteorites. One is that the orbits of meteorites and most comets are greatly dissimilar. Another is that comets have diameters of less than about 20 km and are too small and of too low density to provide the chemical processing required to form the highly differentiated meteorites. Thus, if any meteorites did come from comets, they must be chondritic. Finally, the lifetime of a comet transiting the inner Solar System is only about 10,000 years, whereas exposure ages of meteorites based on the effects of cosmic rays show that all meteorites tested were in space much longer. Thus, it is improbable that any of the analyzed meteorites came from a presently active comet.

The Ages of Meteorites

Various types of radiometric ages can be measured for meteorites. These include: (1) cosmic ray exposure ages, which measure residence times in space; (2) degassing ages, which indicate the time of the last major impact heating; (3) metamorphic ages, which date the time of major reheating and metamorphism within the parent body; and (4) crystallization ages, which date the time of crystallization from a rock melt. For the purposes of this book, the crystallization ages are the most interesting because they date the time that the meteorite first formed as part of a parent body in the Solar System.

Many meteorites, however, were heated and metamorphosed after crystallization in their parent bodies and also have been subjected to subsequent brecciation as well as shock heating by collision events. In these meteorites, original crystallization ages may have been partly or wholly modified. As it turns out, however, the period of time separating crystallization and planetary metamorphism for most meteorites was very short—less than 100 million years—so the ages of metamorphism represent events only slightly younger than the age of the Solar System. Ages for meteorites have been determined by the Rb-Sr, Sm-Nd, Lu-Hf, Re-Os, Ar-Ar, U-Pb, and Pb-Pb methods. Discussion of the results of U-Pb and Pb-Pb measurements will be deferred until the next chapter because they are so important and unique that they deserve a chapter of their own.

As with rocks from Earth and the Moon, there are two approaches to the radiometric dating of meteorites. One involves the analysis of different phases separated from the same meteorite, and it results in an age for that particular meteorite (Table 7.2). The other involves the analysis of whole-rock samples of a number of meteorites of the same type. This sort of analysis results in an age for the particular meteorite type (Table 7.3).

More than 100 meteorites have been analyzed, with the intent of determining the radiometric ages of individual meteorites. Of these, most have yielded Rb-Sr or Sm-Nd isochrons or Ar-Ar age spectra thought to indicate the time of initial crystallization or of a subsequent metamorphism. The majority of these ages fall between 4.4 and 4.6 Ga. There are some meteorites with apparent radiometric ages between 3.4 and 4.4 Ga and a few with ages still younger, but these meteorites invariably show evidence of severe shock heating and metamorphism, and their isotope clocks have likely been wholly or partially reset by these later events. Thus, most of the younger ages are thought to be a consequence of impact phenomena that occurred long after initial crystallization.

Only the meteorites from Mars appear to be truly young. Most of the dated Mars meteorites have radiometric ages of 1–1.5 Ga, but at least one has an age of only 150 Ma. These ages probably record the time of crystallization of particular lava flows on Mars.

Because of their scarcity, small size, paucity of datable material, and fragile nature, carbonaceous chondrites have proven very difficult to date. Several attempts to determine internal Rb-Sr isochrons for these meteorites have not been successful because the data scatter rather than fall on an isochron. Some inclusions from Allende, a brecciated carbonaceous chon-

Table 7.2
Examples of Meteorites with Radiometric Ages That Exceed 4.4 Billion Years

Meteorite		Dating	Age
Name	Type[a]	Method	(billion years)
CHONDRITES			
Allende	Carbonaceous	Ar-Ar	4.52 ± 0.02
		Ar-Ar	4.55 ± 0.03
		Ar-Ar	4.56 ± 0.05
ALH-761	Ordinary	Ar-Ar	4.49 ± 0.10
Bjurbole	Ordinary	Ar-Ar	4.51 ± 0.08
Guarena	Ordinary	Rb-Sr	4.46 ± 0.08
		Ar-Ar	4.44 ± 0.06
Krahenberg	Ordinary	Rb-Sr	4.60 ± 0.03
Menow	Ordinary	Ar-Ar	4.48 ± 0.06
Olivenza	Ordinary	Rb-Sr	4.53 ± 0.16
		Ar-Ar	4.49 ± 0.06
Queen's Mercy	Ordinary	Ar-Ar	4.49 ± 0.06
St. Severin	Ordinary	Sm-Nd	4.55 ± 0.33
		Rb-Sr	4.51 ± 0.15
		Ar-Ar	4.43 ± 0.04
		Re-Os	4.68 ± 0.15
Tieschitz	Ordinary	Rb-Sr	4.52 ± 0.05
Abee	Enstatite	Ar-Ar	4.52 ± 0.03
Indarch	Enstatite	Rb-Sr	4.46 ± 0.08
		Rb-Sr	4.39 ± 0.04
ACHONDRITES			
Bholghati	Basaltic	Rb-Sr	4.54 ± 0.06
		Sm-Nd	4.51 ± 0.03
Ibitira	Basaltic	Sm-Nd	4.46 ± 0.02
		Ar-Ar	4.50 ± 0.03
Juvinas	Basaltic	Sm-Nd	4.56 ± 0.08
		Rb-Sr	4.50 ± 0.07
Moore County	Basaltic	Sm-Nd	4.46 ± 0.05
Stannern	Basaltic	Sm-Nd	4.48 ± 0.07
Y-75011	Basaltic inclusion	Rb-Sr	4.50 ± 0.05
		Sm-Nd	4.52 ± 0.16

(continued on following page)

Table 7.2 (continued)

Meteorite		Dating	Age
Name	Type[a]	Method	(billion years)
	Basaltic matrix	Rb-Sr	4.46 ± 0.06
		Sm-Nd	4.52 ± 0.33
Acapulco	Other	Sm-Nd	4.60 ± 0.03
		Ar-Ar	4.51 ± 0.02
		Ar-Ar	4.51 ± 0.01
		Ar-Ar	4.50 ± 0.03
		Ar-Ar	4.51 ± 0.02
Angra dos Reis	Other	Sm-Nd	4.55 ± 0.04
Lewis Cliff	Other	Sm-Nd	4.55 ± 0.03
STONY IRONS			
Estherville		Rb-Sr	4.54 ± 0.20
		Sm-Nd	4.53 ± 0.09
Morristown		Sm-Nd	4.47 ± 0.02
IRONS			
Caddo County		Sm-Nd	4.53 ± 0.02
		Ar-Ar	4.52 ± 0.01
Colomera		Rb-Sr	4.51 ± 0.04
		Ar-Ar	4.47 ± 0.01
Mundrabilla		Ar-Ar	4.57 ± 0.06
		Ar-Ar	4.54 ± 0.04
		Ar-Ar	4.50 ± 0.04
Pitts		Ar-Ar	4.54 ± 0.06
Sombrerete		Ar-Ar	4.54 ± 0.01
Techado		Ar-Ar	4.49 ± 0.01
Weekeroo Station		Rb-Sr	4.39 ± 0.07
		Rb-Sr	4.28 ± 0.23
		Ar-Ar	4.54 ± 0.03
		Ar-Ar	4.49 ± 0.03

NOTE: The ages are based on either internal isochron (Rb-Sr, Sm-Nd, Re-Os) or Ar-Ar age spectrum methods.

[a] See Table 7.1.

SOURCE: McCoy et al. 1996, Bogard and Garrison 1995, Prinzhofer et al. 1992, Brouxel and Tatsumoto 1991, Nyquist et al. 1990, Lugmair and Galer 1992, Tera et al. 1997, Pellas et al. 1997, Bogard et al. 2000, Takeda et al. 2000, Chen et al. 1998, Stewart et al. 1996, Renne 2000; compilation in Dalrymple 1991.

Table 7.3

Examples of Whole-Rock Isochron Ages of Some Meteorite Types

Meteorite Type[a]	Number of Meteorites Included in Analysis	Dating Method	Age (billion years)
CHONDRITES			
Carbonaceous	4	Rb-Sr	4.37 ± 0.34
Ordinary	17	Rb-Sr	4.52 ± 0.04
Ordinary	6	Rb-Sr	4.44 ± 0.12
Ordinary	13	Rb-Sr	4.49 ± 0.02
Ordinary and enstatite	38	Rb-Sr	4.50 ± 0.02
Ordinary and enstatite	50	Rb-Sr	4.43 ± 0.04
Enstatite	8	Rb-Sr	4.51 ± 0.04
ACHONDRITES			
Basaltic	23	Rb-Sr	4.53 ± 0.19
Basaltic	13	Lu-Hf	4.57 ± 0.19
Other	5	Rb-Sr	4.45 ± 0.18
STONY IRONS	6	Re-Os	4.60 ± 0.05
IRONS	19	Re-Os	4.65 ± 0.11
	16	Re-Os	4.62 ± 0.02
	7	Re-Os	4.60 ± 0.03
	8	Re-Os	4.56 ± 0.01
	10	Re-Os	4.46 ± 0.03
Irons + St. Severin	8	Re-Os	4.57 ± 0.21

NOTE: Some of these types have also been dated by the Pb-Pb method (see Table 8.2).

[a] See Table 7.1.

SOURCE: Shen et al. 1996, 1998; Morgan et al. 1992; Horan et al. 1992; Smolier et al. 1996; compilation in Dalrymple 1991.

drite, have provided a number of Ar-Ar age spectrum ages ranging from 4.52 to 4.56 Ga. These ages probably record a time of major resetting of the K-Ar isotope clock early in the history of Allende. The exact nature of this event is not known, but the likely possibilities include the shock event that formed the Allende breccia or metamorphism of the individual inclusions within the Allende parent body.

Only one whole-rock isochron for carbonaceous chondrites has been obtained (see Table 7.3), but the resulting age of 4.37 Ga using four meteorites has a large uncertainty and therefore is not inconsistent with the Ar-Ar ages measured for Allende. Many carbonaceous chondrites, however, do fall on or very near the whole-rock isochrons formed by analyses of other types of chondrites. Thus, there is little reason to doubt that the carbonaceous chondrites are about 4.5–4.6 Ga in age.

In contrast to the carbonaceous chondrites, the ordinary and enstatite chondrites have provided a wealth of radiometric age data, in the form of both internal isochrons and age spectra for individual meteorites, and whole-rock isochrons for the different chondrite types (Figure 7.5; see also Tables 7.2, 7.3; Figures 4.5, 4.6). The majority of well-dated ordinary chondrites have Rb-Sr and Ar-Ar ages between 4.45 and 4.55 Ga. Because of their low content of rare earth elements and metals, chondrite meteorites are very difficult to date with the Sm-Nd and Re-Os methods, so such ages for these meteorites are rare. Where available, however, they confirm the Rb-Sr and Ar-Ar results.

Most of the ordinary chondrites have been dated by only a single method, primarily Ar-Ar, but there are some exceptions. For example, Guarena and Olivenza give concordant Rb-Sr and Ar-Ar ages. St. Severin has been dated by Rb-Sr, Sm-Nd, Re-Os, Pb-Pb, and Ar-Ar methods, and the ages obtained with these different radiometric clocks are not different within analytical errors.

The data indicate that there are no obvious age differences either between the meteorites of the various chondrite types or between ages measured by different methods. All the radiometric clocks appear to record events about 4.45–4.55 Ga ago. But was this event the initial condensation and aggregation or a subsequent metamorphism? Probably the latter, because most of the dated meteorites are mildly to highly metamorphosed, and it is difficult to see how such changes would not affect the isotope clocks. It is very likely that the Ar-Ar clock would be reset by the metamorphism because it is more susceptible to thermal resetting than the other methods. Thus, it appears probable that the ages for individual ordinary chondrites record metamorphic and subsequent cooling events, perhaps within their parent bodies, rather than the times of condensation and aggregation from the Solar Nebula. The consistency of the individual meteorite ages indicates that these metamorphic events occurred within a relatively short period of time, about 4.4–4.5 Ga ago. The Ar-Ar data on ordinary chondrites whose age spectra show patterns of minimal Ar loss are consistent with a single cooling age (perhaps following metamorphism) of about 4.46 Ga.

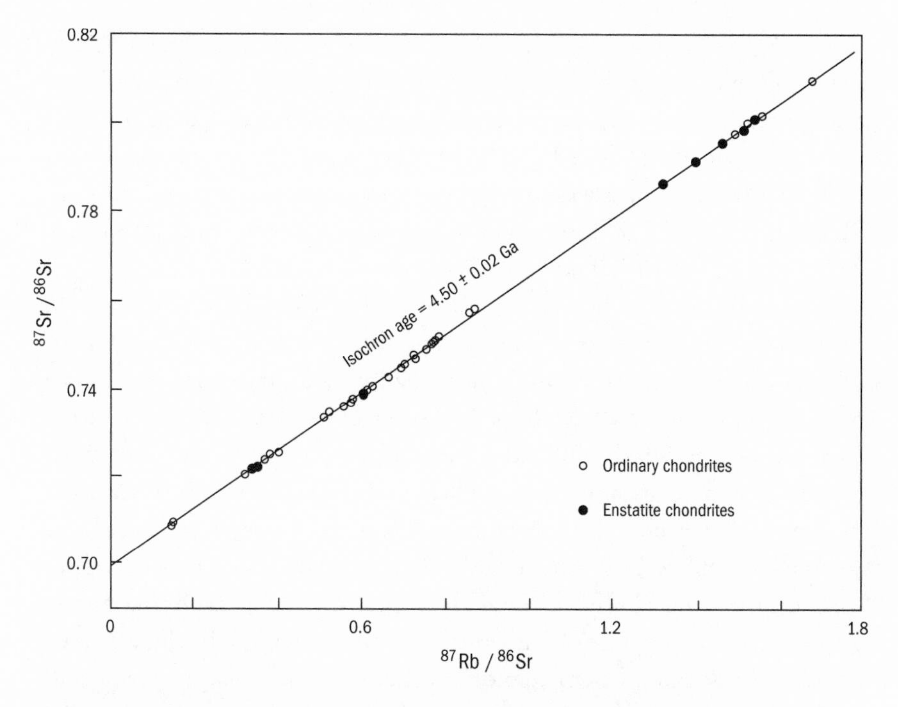

Figure 7.5 A Rb-Sr whole-rock isochron for thirty-eight undisturbed ordinary and enstatite chondrites. (After J.-F. Minster et al., *Nature*, vol. 300, pp. 414–419, © 1982, with permission from Macmillan Magazines Ltd.)

As mentioned above, another way to determine formation ages for meteorites is to measure whole-rock isochrons using a number of meteorites of the same type rather than internal isochrons for single meteorites. Such ages are less likely to have been affected by post-crystallization events than the ages of individual meteorites. A group-by-group Rb-Sr study of chondrites by J.-F. Minster and his colleagues of the University of Paris has shown that, with the exception of a few highly disturbed specimens, the H (high iron and nickel) and LL (very low iron and nickel) ordinary chondrites, and the enstatite chondrites, defined precise isochrons with indicated ages of:

17 H ordinary chondrites	4.52 ± 0.04 Ga
13 LL ordinary chondrites	4.49 ± 0.02 Ga
8 enstatite chondrites	4.51 ± 0.04 Ga

Because these ages are not different within the analytical errors and because the initial ratios are also identical, these three groups of chondrites can be treated as a single isotope system with a common age. The resulting isochron age for all thirty-eight chondrites is 4.50 Ga, an age that probably represents the approximate time of condensation and aggregation of the parent asteroids from the Solar Nebula (see Figure 7.5). The precision with which all these specimens fit the same isochron is quite remarkable, indicating that the ordinary and enstatite chondrite parent bodies formed either simultaneously or within a period of 10–20 million years or less.

There are fewer radiometric ages on individual achondrites than on chondrites, mostly because far fewer achondrites have been found. Of the achondrite meteorites for which high-quality radiometric data have been obtained, nearly half have Rb-Sr or Ar-Ar ages of less than 4.4 Ga. These relatively low ages for the individual achondrites probably represent post-formation impact metamorphism, which is consistent with the observation that most of the achondrites have been brecciated by impacts. Indeed, the Ar-Ar ages of the basaltic achondrites indicate that an impact or series of impacts on the parent body occurred about 3.4–4.1 Ga ago. On the whole, the Sm-Nd isochron ages of individual achondrites tend to be slightly older than the Rb-Sr ages, thereby reinforcing the conclusion that the younger of the Rb-Sr ages are the result of post-crystallization impact events. The older values obtained by the Sm-Nd method are not surprising because the Sm-Nd isotope system tends to be more resistant to post-formational heating than either the Rb-Sr or the Ar-Ar clock.

Most of the older, well-dated achondrites are basaltic, and it is likely that the Sm-Nd ages represent the approximate times of eruption and cooling of the basalt lava on the parent body. This supposition is reinforced by the Rb-Sr and Lu-Hf isochrons on whole rock samples of the basaltic achondrites (see Table 7.3), which are generally concordant with the Sm-Nd ages of individual basaltic achondrites.

The Sm-Nd ages for individual basaltic achondrites and the Rb-Sr and Lu-Hf whole-rock isochrons for these meteorites show that the basaltic achondrites formed about 4.50–4.55 Ga ago, and that their radiometric ages are indistinguishable from those determined for the chondrites. Because the achondrites are the products of magma formation and crystallization, the results indicate that melting and eruption occurred very shortly after formation of the achondrite parent bodies, perhaps within a period of time as short as a few million years.

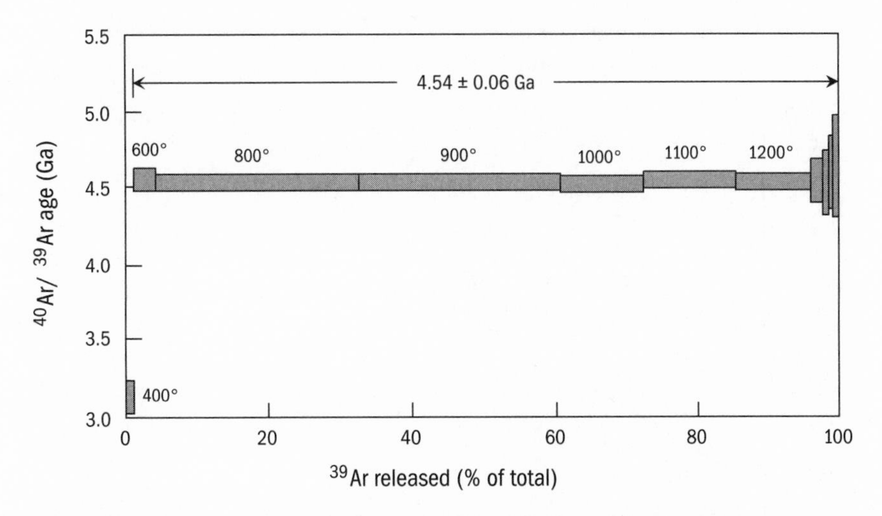

Figure 7.6 An Ar-Ar age spectrum for silicate inclusions within the Pitts iron meteorite. The temperature at which each gas increment was extracted, in degrees Celsius, is shown for each increment. (After S. Niemeyer, *Geochimica et Cosmochimica Acta*, vol. 43, pp. 1829–1840, © 1979, with permission from Pergamon Press plc.)

Within the last few years it has become possible, with improved techniques, to measure precise radiometric ages for iron and stony iron meteorites. Stony iron and some iron meteorites contain small inclusions of silicate minerals, and some of these can be dated by Ar-Ar, Rb-Sr, and Sm-Nd methods. Such ages tend to fall within the range of 4.45–4.55 Ga, although a few have ages as low as 3.8 Ga (Figure 7.6; see also Table 7.2). This could mean that there were two distinct formation times for iron meteorite parent bodies, one very early in the history of the Solar System and another about 3.8 Ga ago. A more likely explanation, however, is that the isotope clocks in the few younger iron meteorites may have been reset by the collision and destruction of their parent body about 3.8 Ga ago.

The Re-Os method can be applied directly to the metallic phases of iron and stony iron meteorites, and there are a number of whole-rock Re-Os isochron ages for these objects (Figure 7.7; see also Table 7.3). The results are similar to the radiometric ages found for individual irons. Metal inclusions in the St. Severin ordinary chondrite also fall on the Re-Os isochron for iron meteorites, indicating that the irons and the chondrites, at least as

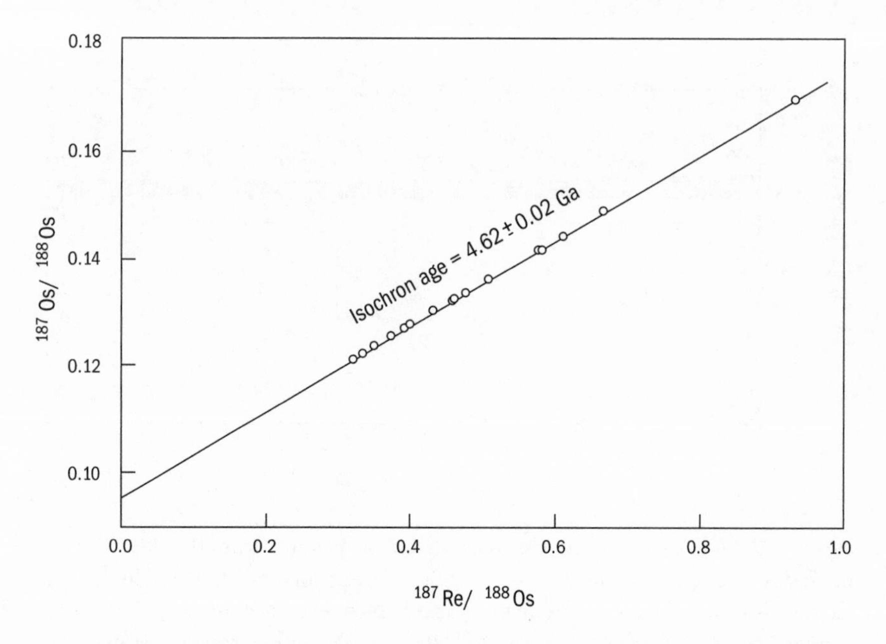

Figure 7.7 Re-Os isochron for sixteen iron meteorites. (After J.J. Shen et al., *Geochimica et Cosmochimica Acta,* vol. 60, pp. 2887–2900, © 1996, with permission from Elsevier Science.)

represented by St. Severin, formed at the same time and from parent material with the same initial osmium isotope composition.

How Old Are Meteorites?

The radiometric data show that there are many meteorites with ages of 4.5–4.6 Ga. Some of these ages, particularly those of the chondrites, represent the formation times of primitive asteroids shortly after their materials condensed and crystallized from the Solar Nebula. Other ages represent the time of differentiation and the formation of metal cores in larger asteroids (stony irons and irons), while others represent the time of eruption of basalt lava flows onto the surfaces of the larger asteroids (basaltic achondrites). The radiometric ages of the oldest iron meteorites are indistinguishable from the ages of the oldest chondrites and achondrites, suggesting that the material of most meteorite parent bodies formed within a period of less than 100 million years about 4.5–4.6 Ga ago.

But is there a "best" age for meteorite formation? Is there some way to connect the ages of the meteorites with the age of Earth? Is there some method of distinguishing between the formation times of the different types of meteorites and thereby learning more details about the early history of the Solar System? The answers to these questions are all yes, and we will explore them in the next two chapters.

CHAPTER EIGHT

Lead Isotopes: Hourglass of the Solar System

Lead isotopes and their bearing on the ages of meteorites and Earth are so important that they deserve a chapter of their own. The fact that two different lead isotopes (^{206}Pb and ^{207}Pb) are produced by the decay of two different uranium isotopes (^{238}U and ^{235}U), along with the very high precision with which lead isotope ratios can be measured, endows the Pb-Pb method with unique and extremely useful properties.

Ask any geologist how old Earth is, and the answer will very likely be close to 4.5 Ga. This number comes partly from the radiometric dating of lunar rocks and meteorites and partly from a model that describes the evolution of lead isotopes in meteorites, Earth, and the Solar System. The foundation of the model was first published in the Russian scientific literature in 1942 by E.K. Gerling of the Radium Institute of the Academy of Sciences of the USSR. Arthur Holmes of the University of Edinburgh and Fritz G. Houtermans of the University of Göttingen, who were unaware of Gerling's work as well as each other's, proposed the same model in 1946. Although neither Gerling, Holmes, nor Houtermans was successful in producing a valid age for Earth, their general approach, usually known as the Holmes-Houtermans model (but should include Gerling), was refined by later researchers. Their approach provides a value that scientists generally accept as the age of Earth, or at least of some event early in the history of Earth's formation.

The Growth of Lead Isotopes over Time

In Chapter 4, we discussed the Pb-Pb isochron diagram briefly without explaining exactly why or how this isotope system produced isochrons. Now we'll look into the way lead isotopes work in more detail, because some of the best evidence for the age of Earth is based on lead isotopes in both meteorites and our planet.

Imagine a system that contains both uranium and lead when it forms. This system can be a mineral grain, a meteorite, a planet, or any discrete volume of rock for that matter. Over time, the lead isotope composition of the system will change because of the radioactive decay of the two uranium isotopes ^{235}U and ^{238}U. This change in lead isotope composition will follow a growth curve (Figure 8.1a). It is curved because the two uranium isotopes decay at different rates and so the lead isotopes they produce, ^{207}Pb and ^{206}Pb, will increase at different rates. As with all isotope methods, the data are normalized by dividing the daughter isotopes by another isotope of lead, ^{204}Pb, whose amount does not change over time because it is neither radioactive nor the product of radioactive decay. The growth curve begins at some initial lead isotope composition. Time is measured along the growth curve, with today being zero years ago.

If the system is closed to uranium and lead, then the exact shape and position of the growth curve on a Pb-Pb diagram will be a function of three quantities:

—The initial lead composition, which defines the starting point.
—The ratio of uranium to lead, which controls how fast the lead ratios grow.
—The length of time the system (rock, meteorite, planet) has been in existence as a distinct and separate uranium-lead system.

Now consider three somewhat different systems—let's say they are meteorites—that originated at the same time with the same initial lead isotope composition but with different ratios of uranium to lead (Figure 8.1b). As time passes, the lead compositions of each of these meteorites will follow different growth curves because they contain different amounts of uranium, so the lead ratios will grow at different rates. When the lead isotopes in these meteorites are measured today, they plot on an isochron that passes through the initial lead composition. Unlike the Rb-Sr and other isochrons, whose slope increases with increasing age, the slope of a Pb-Pb isochron decreases with increasing age. Note that the age of these three meteorites can be found by measuring their lead isotope compositions as they are today; that information alone is enough to define the isochron, whose slope reveals the age.

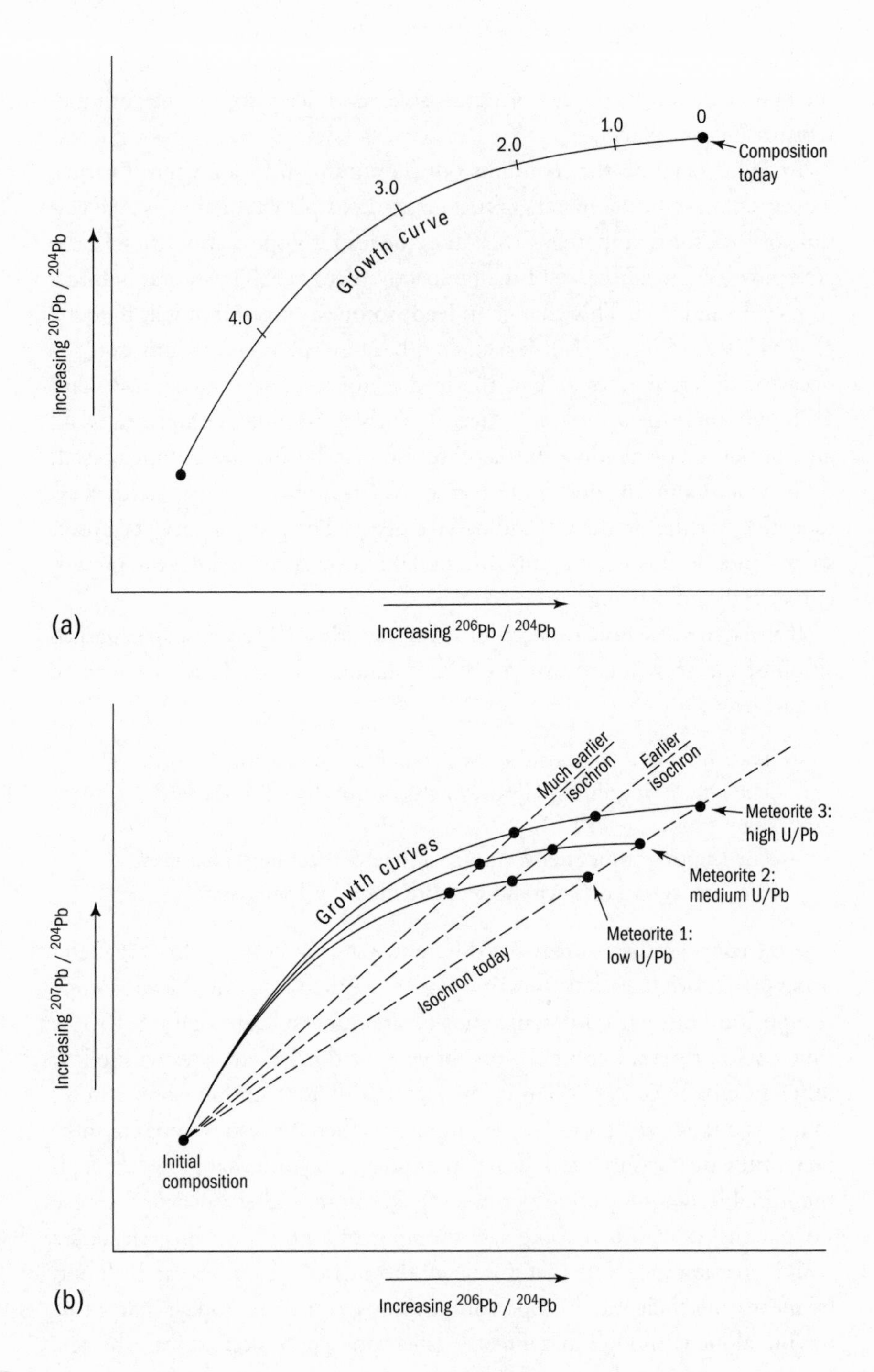

0
1.0
2.0
3.0
4.0
Composition today
Growth curve
Increasing ^{207}Pb / ^{204}Pb
Increasing ^{206}Pb / ^{204}Pb
(a)
Much earlier isochron
Earlier isochron
Meteorite 3: high U/Pb
Meteorite 2: medium U/Pb
Meteorite 1: low U/Pb
Growth curves
Isochron today
Initial composition
Increasing ^{207}Pb / ^{204}Pb
Increasing ^{206}Pb / ^{204}Pb
(b)

It is not necessary to know the initial isotope composition, the ratios of uranium to lead, or the exact shape and position of the growth curves.

To picture the physical situation represented by the growth curves of Figure 8.1b, presume that the initial composition represents the isotope composition of lead, called *primordial lead*, in the Solar Nebula. At that time, discrete bodies (asteroids, proto-planets) were formed, each with the same initial lead isotope composition but with different ratios of uranium to lead. Each growth curve, then, represents the change in lead isotope composition within one of these bodies over time, presuming, of course, that the bodies neither lose nor gain uranium or lead during their history. When the lead isotopes in these bodies are measured today, the isochron defined by the data reveals the age of the bodies.

But the situation can be complicated if lead is removed from a system sometime before today. A system that begins at some initial lead composition and remains on the same growth curve throughout its history is called a single-stage system, and the lead composition that results from that system is a *single-stage lead*. If growth continues to the present, then the lead composition of the system will fall on today's isochron. But perhaps the system was disrupted at some time in the past by geologic processes and some of its lead extracted to form lead ore entirely devoid of uranium. In this instance, the lead isotope composition of the ore will cease to change from that time onward. This lead is still a single-stage lead and it will still plot on its growth curve, but at the time the lead was extracted from the system, not at the zero age point of today. The age represents a "frozen" or fossil record of the lead isotope composition of the system from which the lead ore was extracted at

Figure 8.1 (Opposite) Growth curves for lead isotopes. (a) This growth curve shows the change in lead isotope composition of a uranium-lead system (mineral, meteorite, planet) over time from an initial composition to the composition today due to the decay of radioactive uranium isotopes. Time is measured along the growth curve and labeled in billions of years. (b) These are growth curves for three meteorites with the same initial lead composition but different amounts of uranium relative to lead. The intersections of the three isochrons with the growth curves show the isotope compositions of the meteorites today and at two different times in the past. If the meteorites remain unaltered, then at any time after formation, the lead isotope compositions of the meteorites will fall on an isochron that includes their initial composition and whose slope measures the age of the meteorites.

some specific time in the past. As we'll see later, such frozen leads are important in arriving at a value for the age of Earth or, more specifically, Earth's core.

An alternative to single-stage lead growth is multi-stage growth. If at any time the ratio of uranium to lead in the system is changed, then a new system is created with a new initial lead composition and a new ratio of uranium to lead. This new system will henceforth follow a new growth curve. This type of change may happen any number of times and the leads produced by such systems are called *multi-stage leads.* As you might suspect, multi-stage leads are more difficult, often even impossible, to interpret than single-stage leads. They inherently contain much information about the evolution of Earth's crust and mantle, but the data are very difficult to decipher because there are too many unknown factors. Such leads do not contribute much useful evidence concerning the age of Earth.

The lead isotope systematics represented by Figure 8.1 constitute the Gerling-Holmes-Houtermans lead isotope model, which, one way or another, is the basis for nearly all age of Earth calculations. There are two fundamental difficulties with this model. One is that it requires samples of single-stage leads. Many meteorites meet this requirement, but finding single-stage leads on Earth is not nearly as easy. Another problem is that the model cannot be used to calculate an age for Earth without making some assumptions about the distribution of lead isotopes in the early Solar System, more specifically about the genetic relationship between meteorites and Earth. These important points will be discussed later in the chapter.

With the above discussion as background, let's return to Gerling, Holmes, and Houtermans. We'll see how their early attempts to use lead isotopes to determine the age of Earth's crust eventually led their successors to use the Gerling-Holmes-Houtermans model to measure the age of the meteorites, Earth, and the Solar System.

Gerling, Holmes, Houtermans, and the Age of the Crust

The explosive growth of physics during the early twentieth century resulted in the development of many new instruments to explore the nature of matter and its constituents. One of these instruments was the mass spectrograph, which used a magnetic field to separate nuclides of different mass and photographic plates to estimate their abundance. This clever instrument was a forerunner of the modern mass spectrometer, which uses electronic detectors instead of photographic plates to measure precisely the isotope ratios.

As mentioned in Chapter 4, James J. Thomson of the Cavendish Laboratories at Cambridge University first measured isotopes and proved their existence. Within a few short years, F.W. Aston, working in Thomson's laboratory, redesigned Thomson's parabola mass analyzer and set about determining the isotopes of a variety of elements. In 1929, he measured the lead isotope composition of a sample of uranium ore and found that it was greatly enriched in ^{206}Pb relative to ^{207}Pb (primarily because ^{238}U is much more abundant than ^{235}U). From Aston's data, C.N. Fenner and C.S. Piggot of the Carnegie Institution's Geophysical Laboratory in Washington, D.C., calculated the first isotope age based on the decay of ^{238}U to ^{206}Pb.

The first isotope ages based on the ratio of $^{207}Pb/^{206}Pb$ were published in 1936 by J.L. Rose and R.K. Stranathan of New York University, who pointed out that this ratio must vary systematically with time because ^{235}U and ^{238}U decay at different rates. During the years 1938–1941, Alfred O. Nier and his colleagues at the University of Minnesota published several papers in which they reported systematic variations in the proportions of ^{206}Pb and ^{207}Pb relative to ^{204}Pb in uranium and lead ores. They proposed that these variations were due to a mixture of primordial lead and radiogenic lead, the latter of which was a function of geologic time. By 1941, the way had been prepared to estimate the age of Earth based on new principles and new isotope data, and Gerling quickly seized the opportunity.

Nier and his coworkers had found one lead ore sample, a lead sulfide mineral called galena from Ivigtut, Greenland, whose $^{206}Pb/^{204}Pb$ and $^{207}Pb/^{204}Pb$ ratios were extremely low, and he speculated that the amount of radiogenic lead in this sample was small or negligible. Gerling used the lead isotope ratios in the Ivigtut sample as primordial for the purposes of his calculations, and subtracted these values from the lead ratios of seven young lead ores whose geologic ages were known. In this way he found the radiogenic lead fraction in each of the young ores. Using the average radiogenic $^{207}Pb/^{206}Pb$ of the seven ores and the average age of the ores (130 Ma), Gerling calculated an age for Earth of 3.23 Ga, which he regarded as a minimum estimate.

Gerling's calculation is shown graphically in Figure 8.2. He had, in effect, determined a two-point isochron using the Ivigtut analysis as one point (the initial lead) and the average of the seven lead ores for the other point. The slope of the isochron was equivalent to an age of 3.1 Ga, which was the av-

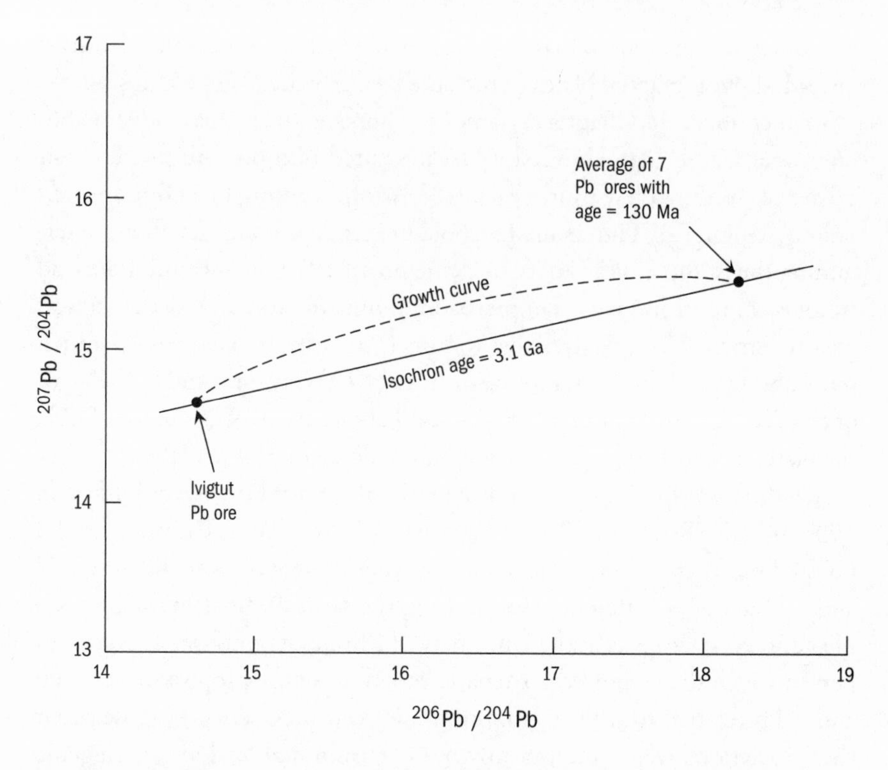

Figure 8.2 Gerling's minimum age for Earth. Gerling used the Ivigtut lead ore (the mineral galena–lead sulfide) to represent primordial lead. He added the average age of the seven young lead ores (130 Ma or 0.13 Ga) to the isochron age (3.1 Ga) to obtain a minimum age of Earth of 3.23 Ga.

erage age of the seven ores when they formed, about 130 million years ago (0.13 Ga). The two numbers added together give 3.23 Ga.

Gerling went through the same exercise comparing a galena from Great Bear Lake, whose age was 1.25 Ga, to the Ivigtut ore. This calculation yielded a minimum age for Earth of 3.95 Ga. From these computations, he concluded that Earth's age is no less than 3–4 Ga.

Several assumptions are implicit in Gerling's calculation. Foremost are the presumptions that (1) the seven lead ores originated from the same homogeneous source whose initial lead isotope composition was identical to the Ivigtut galena, and (2) all of the leads are single-stage leads. Gerling's results, as he fully realized, are minimum values for the age of Earth. While of the correct order of magnitude, they are too low primarily because the lead isotopes in the Igvitut galena are not of primordial composition, but also because the young lead ores are not single-stage leads. Nonetheless, he had devised a fruitful approach that others would eventually exploit to the fullest.

Arthur Holmes developed his ideas independent of Gerling—Gerling's paper was buried in the Russian scientific literature—and presented them in a series of papers published between 1946 and 1950. Holmes's approach was considerably more cumbersome than Gerling's, and I will not explain it in detail. But his method consisted, in effect, of extrapolating pairs of Pb-Pb isochrons back to a common intersection to find their initial lead composition and age, although Holmes did not explain or actually do it in this way. Holmes's data set consisted of lead isotope data obtained by Nier and his colleagues on 25 lead ores, and from these he found 1419 solutions for age. His solutions ranged from 2 Ga to more than 4 Ga, with a pronounced concentration at 3.35 Ga. Holmes concluded that the most probable age of Earth is about 3350 Ma.

Holmes's age for Earth is invalid because the samples do not conform to the assumptions of the method. Holmes assumed that the lead ores evolved in separate systems within the crust, that each system had a distinct ratio of uranium to lead, that all the systems originated at the same time and with the same primordial lead composition, and that all the leads were single-stage leads. If these assumptions were true, then Holmes's method would have yielded the same solution for the age of Earth for all pairs of lead ores. The wide range of his results, however, clearly shows that the twenty-five leads he used do not meet the conditions required by the method.

Houtermans used a slightly different approach than either Holmes or Gerling, although the basic method was similar, as was his use of the lead data of Nier and his colleagues. Houtermans pointed out that on a plot of $^{207}Pb/^{204}Pb$ versus $^{206}Pb^{204}/Pb$, lead minerals of the same age must lie on straight lines, which he called "isochrones" (isochrons), whose slope is a function of the time of origin and the time the lead was separated from its uranium. Moreover, said Houtermans, two or more isochrons must intersect at the primordial (initial) composition and give a unique solution for the age of Earth. Using Nier's data, Houtermans found an age of 2.9 ± 0.3 Ga. This age he called "the age of uranium." Houtermans thought that it represented the age of the Solar System, that is, the time of formation of terrestrial uranium, provided there was not enrichment of uranium relative to lead during formation of

Earth's crust. If such enrichment occurred, and he cited some evidence that it had, then the age he found was the age of the crust.

Houtermans' method was based on the same assumptions as Holmes's and suffered from the same flaws, but he had advanced the final concept in the Gerling-Holmes-Houtermans model—the concept of lines of equal time—and had given isochrons their name. In addition, he cleverly suggested that a better value for the composition of primordial lead might be found by analyzing iron meteorites.

Claire C. "Pat" Patterson of the California Institute of Technology spent most of his scientific career measuring lead in the environment. He deserves much of the credit for convincing policy makers that the world would be a healthier place if paint, gasoline, and other common substances were lead-free. His scientific fame, however, rests largely on his earlier work on the age of Earth. In 1953, he and his colleagues measured the lead isotope composition and the amount of uranium and lead in the mineral troilite (a form of iron sulfide) from the iron meteorite Canyon Diablo, which excavated Meteor Crater (see Figure 7.1, top). The troilite contained the lowest ratios of ^{206}Pb and ^{207}Pb to ^{204}Pb ever measured and was also exceedingly low in uranium relative to lead. The low ratio of uranium to lead meant that the lead isotope composition could not have changed significantly since the meteorite was formed. Thus, suggested Patterson and his colleagues, the lead ratios in the troilite might be primordial lead.

Houtermans was quick to take advantage of the new lead data for Canyon Diablo. In the same year (1953), he published a paper in which he calculated an age for Earth that is very close to the current accepted value. He made two principal assumptions. The first was that the isotope composition of lead at the time of the formation of Earth's crust was represented by the values Patterson had found in the troilite of Canyon Diablo. The second was that the majority of the measured Tertiary (era) leads had evolved by single-stage growth beginning at a common time of origin up to the time of formation of the lead ores in which they occur. He chose lead ores of Tertiary age because the calculated age of Earth is relatively insensitive to errors in the geologic age of the ores when they are young (but is more sensitive to the single-stage assumption).

In 1953, the literature contained lead isotope measurements for twenty-two lead ores, from which Houtermans selected ten of the youngest. For each ore he calculated the slope of a two-point isochron through the ore and

Canyon Diablo troilite. He used the average of the ten slopes to calculate the age of Earth as 4.5 ± 0.3 Ga. Graphically, Houtermans' solution is nearly identical to Gerling's (see Figure 8.2), except that he used the Canyon Diablo troilite instead of the Ivigtut galena for the initial lead isotope composition and the average of the isotope compositions of ten young lead ores instead of seven.

Houtermans' 1953 result is notable not only because it is near today's value, but also because it was the first calculation to link the age of Earth to the age of meteorites, thereby implying that their origins were related. Houtermans made no attempt to justify the assumption that Earth and meteorites were part of the same lead system, but, as will be shown later, a reasonable case can be made for its validity.

Thus, in 1953 there were two somewhat different ways of applying the Gerling-Holmes-Houtermans model to the problem of the age of Earth. One, the "ore method" of Gerling, used the change in lead isotopes between the times of formation of an ancient lead ore and a recent one. This method gave ages of about 3.3 Ga that were really minimum ages for Earth. The second method, the "meteorite method" of Houtermans, consisted of using the lead isotope composition in the troilite phase of iron meteorites for the primordial lead composition and calculating the time required to form lead with the isotope composition of young lead ores. This second method required the assumption that the lead in iron meteorites and the lead of Earth shared a common time and initial composition, an assumption that had yet to be proved. Evaluating the situation in 1955, Patterson and his colleagues concluded that while there was some reason to think that there was a connection between meteorites and Earth, results based on such an assumption should be viewed with considerable skepticism.

Patterson and the Meteoritic Lead Geochron

Houtermans had assumed there was a connection between meteorite lead and terrestrial lead, but he had not provided any arguments for the validity of this assumption. Moreover, Houtermans' isochron was based on only two points: troilite lead from the Canyon Diablo meteorite and the average lead in Tertiary ores. Patterson corrected both deficiencies in a classic 1956 paper. Using the lead isotope analyses of three stone meteorites and the troilite phase of two iron meteorites, he showed that these data formed an isochron

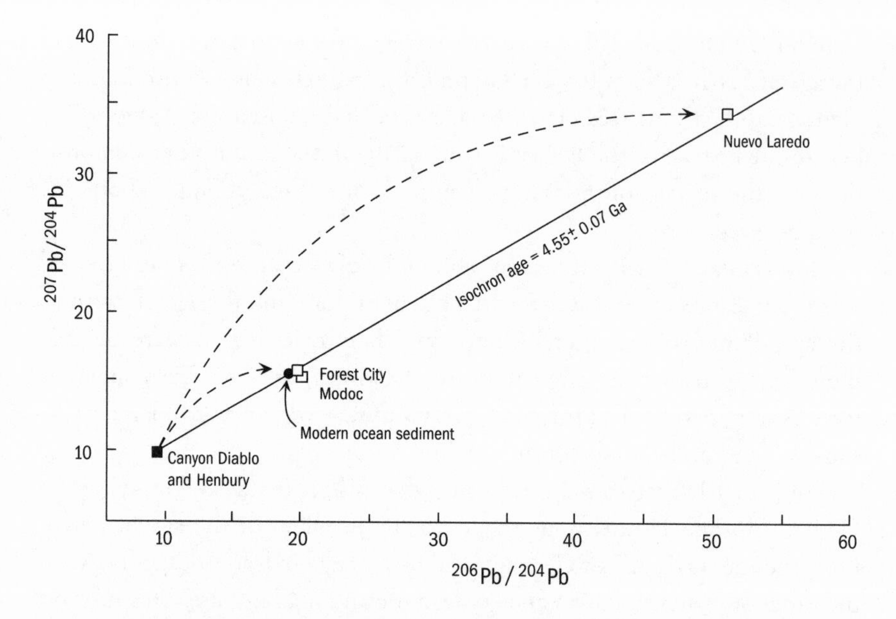

Figure 8.3 Patterson's meteorite lead isochron. This analysis used three stone meteorites (open squares) and two iron meteorites (solid square). The lead composition of modern ocean sediment (solid circle) falls on the isochron, suggesting that meteorites and Earth are related and of the same age. The dashed lines are growth curves. (After C.C. Patterson, *Geochimica et Cosmochimica Acta*, vol. 10, pp. 230–237, © 1956, with permission of Pergamon Press plc.)

(Figure 8.3) whose slope was equivalent to an age of 4.55 ± 0.07 Ga using the decay and abundance constants commonly employed at that time. (The calculated age for these same data is 4.48 Ga using today's more accurate constants.) Such linearity from a set of data with a wide range in isotope composition, Patterson argued, strongly indicated that these five meteorites fulfilled the assumptions of the Pb-Pb age method. Any meteorite in which uranium and lead had been fractionated would not fall on the isochron. Patterson concluded, therefore, that the isochron age represents the time that meteorites formed.

Patterson's next step was to make the connection between meteorites and Earth. He noted that if modern Earth lead fell on the meteorite isochron, then it too must have evolved in a closed system from an initial lead composition the same as in meteorite troilite over the past 4.55 Ga.

But where is there a representative sample of modern Earth lead? This is not a simple problem, because the crust of Earth has a very complex history.

Patterson proposed that modern sediment from the deep ocean might represent average crustal lead because such sediment samples a wide volume of material from the present continents. When measured, the lead isotope composition of Pacific deep-sea sediment satisfied Patterson's prediction, plotting on the meteorite isochron. Patterson concluded, "independently measured values . . . satisfy [the] expressions . . . and therefore the time since the earth attained its present mass is $4.55 \pm 0.07 \times 10^9$ years" (Patterson 1956, p. 138). Patterson had determined a convincing age for meteorites and had also shown it to be highly probable that Earth was part of the meteorite lead system and of the same age.

Six years later, Patterson teamed with V.R. Murthy of the University of California at San Diego. They refined the age of meteorites and strengthened the hypothesis that Earth was part of the meteorite lead system. Murthy and Patterson selected lead isotope analyses of five stone meteorites thought most likely to have been closed systems since formation. To these data they added the composition of primordial lead, which they took to be the average composition of lead in troilite from five iron meteorites. These data formed an isochron with an age of 4.55 Ga. Because of the unique nature of this isochron in the history of the Solar System, they named it the meteoritic geochron, or simply the *geochron.*

Their next task was to show that the meteoritic geochron also represents the evolution of lead isotopes in Earth, that is, is identical to the terrestrial geochron. To do this, Murthy and Patterson used two data to represent terrestrial lead. One was the average isotope composition of lead in more than 100 samples of recent North Pacific sediment. This lead isotope composition, they reasoned, should lie slightly to the right of the terrestrial geochron because marine sediment is eroded from rocks of the upper layers of the crust, which were enriched in uranium relative to lead by the crustal formation processes. The other was single-stage lead ores that had followed the crustal growth curve and had been extracted from their uranium-bearing source in the past. These should plot to the left of the terrestrial geochron. Murthy and Patterson chose the average lead isotope composition of ores from Bathurst, New Brunswick, which have an age of about 350 Ma and which were at the time, but no longer, thought to be single-stage leads. These two data points, they reasoned, should bracket the position of the terrestrial geochron, and since they also bracketed the meteoritic geochron, the two geochrons must be very nearly the same, if not identical. Moreover, the North Pacific and

Bathurst leads both lie on a growth curve that passes through the primordial meteorite (troilite) lead composition and satisfies what was then known about the average ratio of uranium to lead in the crust. Murthy and Patterson reasoned that these things could not be a coincidence, so they concluded that meteorites and Earth's crust are parts of the same lead isotope system, and that the age of meteorites and the age of Earth are therefore the same.

In addition to refining the lead isotope age of meteorites, Murthy and Patterson provided a sound basis for connecting lead growth in Earth, a body whose time of origin cannot be determined directly, with lead growth in meteorites, whose ages can be precisely measured. Houtermans had assumed that meteorites and Earth originated at the same time as part of the same uranium-lead system. Murthy and Patterson had shown that such an assumption was not only reasonable but probable. Subsequent researchers would improve on this model and leave little doubt that the age of meteorites represents, to a close approximation, the age of Earth, or at least the material from which Earth is made. But before continuing with lead growth in Earth, let's see what lead isotopes say about the ages of individual meteorites.

Pb-Pb Ages of Meteorites

There are not nearly as many Pb-Pb ages on meteorites as there are Ar-Ar and Rb-Sr ages, primarily because precise lead measurements on these objects are rather difficult. One reason for the difficulty is that most meteorites contain only a very small amount of lead, often less than one part per million. Another is the ease with which the samples may become contaminated. Lead is not only a volatile element; it is ubiquitous in today's environment. It occurs, for example, in some older paints, and there are relatively high concentrations of lead in the environment from the exhaust of vehicles that burn fuel containing lead tetraethyl. Because of the problem of contamination, precise lead isotope measurements must be made in special lead-free laboratories, and there are only a few institutions in the world where such facilities exist.

Despite the difficulties, Pb-Pb ages commonly are more precise than ages measured by other radiometric methods, for two reasons. First, the critical measurement is an isotope ratio of the same element (lead) rather than a ratio based on isotopes of two different elements (rubidium and strontium, uranium and lead, etc.), and the former can be made much more precisely than the latter. Second, the $^{207}Pb/^{206}Pb$ pair, consisting of two isotopes of the same

Table 8.1

Examples of Pb-Pb Internal Isochron Ages for Individual Meteorites

Meteorite		Age
Name	Type	(billion years)
Allende	Carbonaceous chondrite	4.553 ± 0.004
		4.565 ± 0.004
Mezo-Madras	Ordinary chondrite	4.480 ± 0.002
Sharps	Ordinary chondrite	4.472 ± 0.005
Barwell	Ordinary chondrite	4.559 ± 0.005
Bruderheim	Ordinary chondrite	4.482 ± 0.017
Appley Bridge	Ordinary chondrite	4.569 ± 0.018
St. Severin	Ordinary chondrite	4.543 ± 0.019
Bovante	Basaltic achondrite	4.510 ± 0.08
Cachari	Basaltic achondrite	4.453 ± 0.030
Juvinas	Basaltic achondrite	4.556 ± 0.012
		4.540 ± 0.001
Moama	Basaltic achondrite	4.426 ± 0.188
Moore County	Basaltic achondrite	4.484 ± 0.038
Nuevo Laredo	Basaltic achondrite	4.514 ± 0.030
Pasamonte	Basaltic achondrite	4.53 ± 0.03
Serra de Mage	Basaltic achondrite	4.399 ± 0.070
Angra dos Reis	Cumulate achondrite	4.544 ± 0.002
Estherville	Stony iron	4.555 ± 0.035

NOTE: All measurements were made since analytical techniques were greatly improved in the mid-1970s.

SOURCE: Brouxel and Tatsumoto 1991, Tera et al. 1997; compilation in Dalrymple 1991.

element, is immune to chemical fractionation during a post-crystallization disturbance, whereas chemically dissimilar pairs of elements are not.

Pb-Pb ages can be measured in three ways: (1) internal isochron ages based on lead isotope measurements of three or more minerals from a single meteorite, (2) whole-rock isochron ages based on measurements on three or more meteorites of the same type, and (3) model ages for individual meteorites based on a composition for primordial lead (usually Canyon Diablo troilite) and a lead isotope measurement on a single meteorite.

There are precise Pb-Pb isochron ages for only a few dozen individual meteorites (Table 8.1). All but a few, which were probably altered by some later metamorphic event, fall within the narrow range of 4.53–4.57 Ga, a

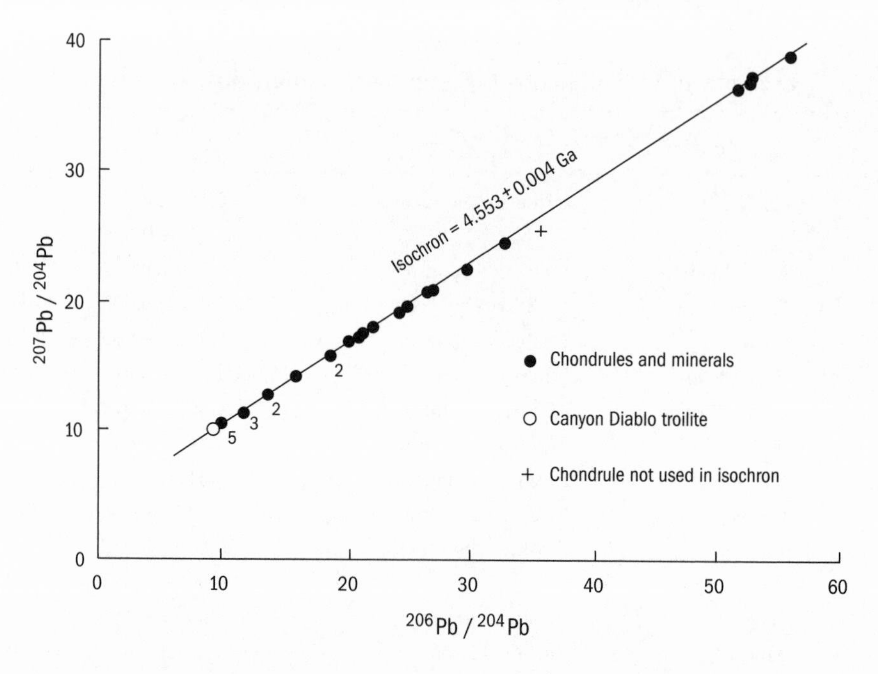

Figure 8.4 A Pb-Pb isochron for the carbonaceous chondrite Allende. This analysis used twenty-eight mineral fractions and chondrules separated from the meteorite (solid circles). The isochron passes through the composition of Canyon Diablo troilite (open circle). Where data are too tightly clustered to be shown individually, the number of data represented by a single symbol is indicated. The data for one chondrule was not used to calculate the isochron age, but the result is not significantly affected by its exclusion. (After Tatsumoto et al., *Geochimica et Cosmochimica Acta,* vol. 40, pp. 617–634, © 1976, with permission from Pergamon Press plc.)

scatter of less than 1%. In addition, there is no obvious difference in the Pb-Pb isochron ages of chondrites and achondrites.

Some of the internal Pb-Pb isochrons are quite impressive. The one for the carbonaceous chondrite Allende, for example, is based on twenty-eight analyses of chondrules, aggregate, and matrix fractions (Figure 8.4). These data fall on a line that passes through the composition of Canyon Diablo troilite and whose slope has an error of less than a tenth of a percent.

The few modern whole rock Pb-Pb isochrons for meteorite types fall within the range 4.54–4.58 Ga (Table 8.2). This range coincides very closely with the isochron ages of individual meteorites. Like the ages for individual meteorites, the isochrons for meteorite types form precise linear arrays that

Table 8.2

Pb-Pb Whole-Rock Isochron Ages of Different Meteorite Types

Type	Number of Meteorites in Analysis	Age (billion years)
Ordinary chondrites	5	4.551 ± 0.007
Ordinary chondrites	4	4.577 ± 0.004
Ordinary chondrites	8	4.554 ± 0.006
Ordinary chondrites	6	4.556
Basaltic achondrites	9 + p	4.540 ± 0.020

NOTES: All measurements were made since techniques were greatly improved in the mid-1970s. p indicates that the primordial Pb ratio, as determined from iron meteorites, was included as a datum in the isochron.

SOURCE: Göpel et al. 1994; compilation in Dalrymple 1991.

pass through the composition of Canyon Diablo troilite, and there are no detectable differences between the ages of chondrites and achondrites.

Model Pb-Pb ages for meteorites are calculated assuming a single-stage history for the meteorite and using the composition of Canyon Diablo troilite for the composition of primordial lead. Each model age is, in effect, a two-point isochron age, the points being the lead isotope compositions of Canyon Diablo troilite and of the meteorite. They are called model ages to distinguish them from isochron ages, the latter being based on three or more data where a straight line relationship does not have to be assumed but is a result of the measurements. Like the internal and whole-rock isochrons, most of the Pb-Pb model ages fall within the relatively small range of 4.52–4.57 Ga, with a pronounced concentration of 4.55–4.56 Ga (Figure 8.5). There are some with values both higher and lower, but the total range of the model ages is only a few percent.

An interesting feature of the Pb-Pb isochron and model ages is that they tend to be somewhat higher, a percent or so, than the Rb-Sr and Ar-Ar ages for meteorites (compare with Tables 7.2 and 7.3). This is probably because the different radiometric systems are measuring slightly different events. Scientists think that the Pb-Pb ages record the time of last homogenization of lead isotopes throughout the Solar Nebula, the Rb-Sr ages the formation times of the parent bodies, and the Ar-Ar ages the times of cooling. These differences, along with metamorphic and shock disturbances on the parent bodies, which would disturb the Pb-Pb ages somewhat less than the Rb-Sr or Ar-Ar ages, probably accounts for the observed age variations. Variations

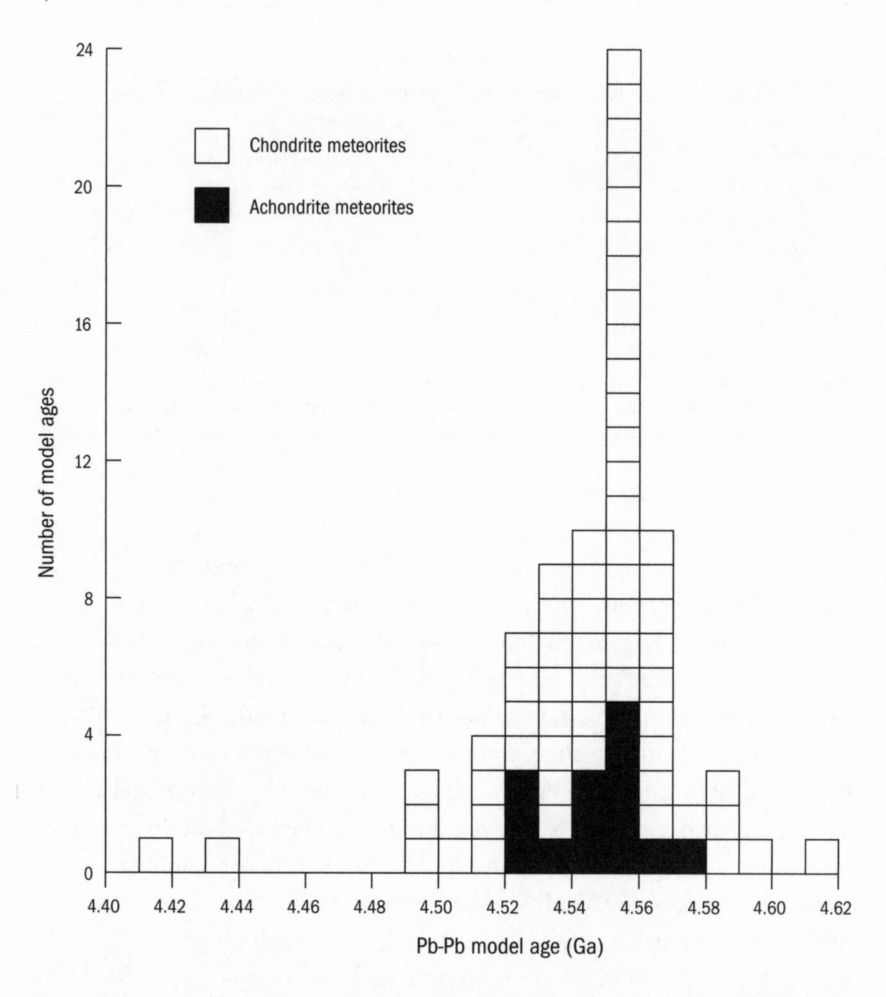

Figure 8.5 Pb-Pb model ages for chondrite and achondrite meteorites. Model ages are essentially two-point isochrons using the lead isotope compositions of the mineral troilite from Canyon Diablo, or another iron meteorite, and the lead isotope composition of the meteorite or a mineral it contains.

in the responses of the different radiometric methods offer the possibility of learning more about the detailed history of events in the early Solar System.

The Pb-Pb ages of meteorites, however measured, form a remarkably consistent set of data that leads to two important conclusions. First, the isochron and model Pb-Pb ages of meteorites are in excellent agreement and average about 4.55 Ga. This age represents, within a small fraction of a percent, the time that the composition of lead isotopes was last uniform throughout the

Solar System, that is, when solid objects first formed. Second, there are no detectable differences between the Pb-Pb ages of chondrites and achondrites. This indicates that condensation from the Solar Nebula, aggregation of the meteorite parent bodies, and the eruption of lava flows onto the surfaces of the achondrite parent bodies occurred within a very short period of time, probably only a few millions of years.

The Terrestrial Connection

There is incontrovertible evidence from lead isotopes that the age of meteorites is approximately 4.55 Ga, an age that is confirmed by the results from the other radiometric dating methods. If all the solid bodies of the Solar System formed nearly simultaneously, then the Pb-Pb age of meteorites also represents the age of Earth. This is a perfectly reasonable conclusion and is probably correct, at least as a good approximation. But, as Patterson recognized nearly five decades ago, it is much more satisfying if it can be demonstrated that Earth is part of the same lead isotope system as meteorites, rather than just assume the relationship, and if a more refined age can be assigned to Earth.

The present lead composition of meteorites can be explained quite simply if meteorites were all formed at 4.55 Ga with the same primordial lead isotope composition but different ratios of uranium to lead. Through time, the lead isotope composition of each meteorite (and each parent body) changed along its own single-stage growth curve to a position today on an isochron of zero age—the geochron.

One approach is to treat Earth like a meteorite and presume that it formed at the same time as the meteorites, with its own ratio of uranium to lead. It then follows that the lead isotope composition of Earth must have evolved along a growth curve that began at the meteorite primordial composition of lead. A problem with this approach is that there is no sample that represents the entire planet—it is simply too big and too complex, and too much of it is inaccessible. A growth curve, however, need not necessarily involve the entire Earth. Any uranium-lead reservoir within Earth, such as the mantle or crust or a portion of either, would also have a growth curve as long as that reservoir formed at the same time as Earth.

The hypothesis of a single growth curve for Earth seems reasonable, because modern sediments and young lead ores plot very close to the meteoritic geochron. But a more convincing case could be made if the terrestrial growth curve actually could be reconstructed, if the curve passed through the meteorite primordial lead isotope composition, and if the age was 4.55 Ga, or thereabouts.

As Gerling recognized more than four decades ago, lead ores represent the isotope composition of lead in their source rocks at the time the ores were extracted from that source. Thus, they represent the fossil lead isotope record of some uranium-lead reservoir within Earth, probably some volume of the lower crust or upper mantle. Isotope analyses of single-stage lead ores should, theoretically, enable scientists to trace the evolution of lead isotopes in Earth through time and to reconstruct the terrestrial growth curve.

One difficulty in implementing this idea involves finding the right samples, because in order to reconstruct the single-stage growth curve for Earth, it is necessary to identify which lead ores are single-stage. Since most lead ores are the products of multi-stage processes, how can single-stage ores be identified? Conformable ores are a possible solution to this problem. *Conformable lead ores,* also called stratiform ores, are so-named because they conform to the geometry of the sedimentary beds within which they are found. They do not intrude or crosscut the sedimentary rocks but occur as bedded deposits within them. They are thought to have formed in the ocean by the deposition of lead sulfide resulting from volcanic eruptions that brought metals directly from the mantle or lower crust to the surface. If this is so, then conformable ores are the same age as the enclosing sedimentary rocks, and they have not been modified or contaminated by multi-stage igneous processes within the crust, as has occurred with other lead ores.

During the 1960s and early 1970s, reconstructing Earth's growth curve was a major goal of lead isotope studies, and many papers were written on the subject. Conformable ores are not numerous, but lead isotope analyses of the dozen or so deposits do fit a single-stage growth curve that passes through the composition of Canyon Diablo troilite. Figure 8.6 shows an example of one such single-stage growth curve based on thirteen conformable lead ores ranging in age from 0.1 Ga to nearly 3.3 Ga. The geochron for this growth curve, however, is not 4.55 Ga but 4.43 Ga. Despite numerous attempts over the years, it is not possible to construct a single-stage growth curve that passes through the conformable ores and through the lead isotope composition of Canyon Diablo troilite, and also has a geochron age of 4.55 Ga. The

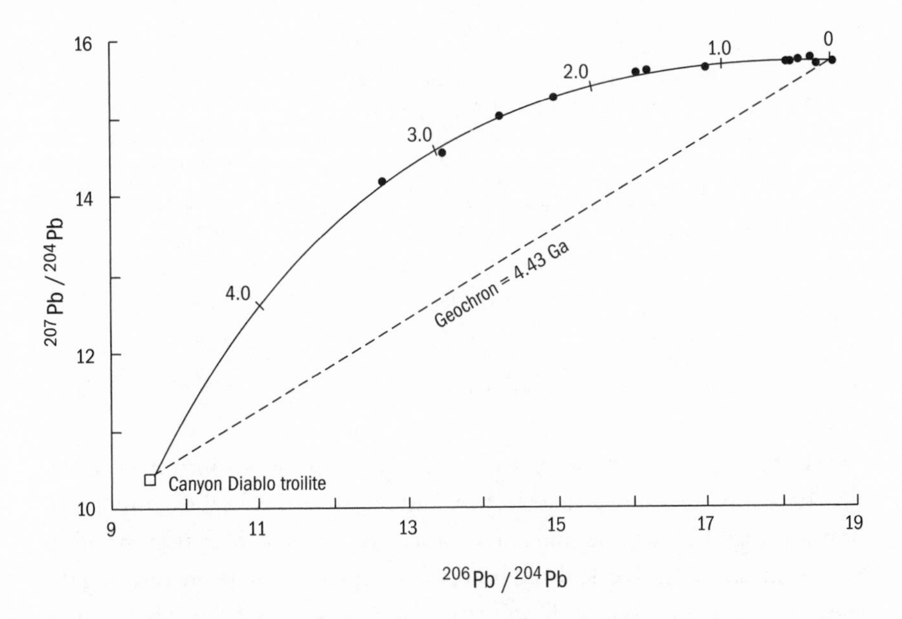

Figure 8.6 Conformable lead ores. The lead isotope compositions of thirteen well-dated conformable lead ores fit a single-stage growth curve with an age of 4.43 Ga for Earth. (After Doe and Stacey 1974.)

conformable lead data fit a single-stage growth curve only if the age of Earth is somewhat less than 4.5 Ga, which does not agree with the measured ages of meteorites.

Another approach to finding the age of Earth is virtually identical to the one developed by Gerling in 1942. Recall that he calculated the length of time required for the isotope composition of a lead ore of known age to evolve from some primordial value to its fossil value at the time the lead ore was separated from its source. Adding this evolution time to the age of the ore then represents the age of Earth, or at least the age of the lead reservoir from which the ore was derived. Gerling had used the Ivigtut galena, which was not primordial in composition, so his result was flawed but his general approach was sound.

Gerling used young (130 Ma) lead ores for his calculations, but a much better approach is to use very old leads, that is, leads older than about 3.5 Ga. The advantage of using ancient leads is that they spent less time than young leads evolving in their lead reservoir, and they did so very early in Earth's history, so they are more likely to be single-stage leads. Even if the ancient leads

Table 8.3
Ages of Earth's Core Calculated from Lead Isotopes in Well-Dated Ancient Lead Ores

Lead Ore	Age of Ore (billion years)	Age of Earth's Core (billion years)
Isua, Greenland	3.807 ± 2	4.492
	3.807 ± 2	4.419
	3.708 ± 3	4.510
Pilbara, Australia	3.465 ± 3	4.495

SOURCE: Data from McCulloch 1996.

turn out not to be single-stage leads, the errors introduced into the calculation by assuming a single-stage history and by any errors in the composition of primordial lead will be small. The other side of the coin is that the accuracy of an age of Earth calculated in this way depends heavily on knowing the ages of the lead ores very well, so the use of only well-dated ores is essential.

Although there are not many well-dated, ancient lead ores, a recent analysis by Malcolm McCulloch of the Australian National University identified four. He calculated ages of Earth (he called them ages of Earth's core, for reasons to be explained shortly) using Canyon Diablo troilite for the composition of primordial lead. His results are shown in Table 8.3.

But McCulloch took his analysis a step further. He reasoned that since the leads do not give ages for Earth equal to chondrite meteorites, then the Canyon Diablo value for primordial lead is incorrect for these ores. The value must be adjusted to account for the change in lead composition that occurred between the time the meteorites formed and the time the lead reservoir for the ancient Earth leads formed. In other words, he used a two-stage model. After making this correction, he found that the ancient leads, as well as the conformable lead ores shown in Figure 8.6, fit a growth curve with an age of Earth of 4.49 Ga.

It appears that the best attempts to reconstruct the terrestrial lead growth curve and to find an age for Earth from ancient lead ores result in an age some 60 million years or so less than the ages of meteorites. Given the precision of the data, this difference is significant, but what does it mean? The most likely explanation is that the age of 4.49 Ga represents some event in Earth's history that occurred after accretion began, such as differentiation of the Earth into core, mantle, and crust. Stephen Galer and Steven Goldstein of the Max Planck Institute for Chemistry in Mainz, Germany, have noted that the lead content of the silicate part of the Earth is lower by a factor of 65 or so—neglecting hydrogen, of course—than the Sun and carbonaceous chondrites. This is most easily explained if lead followed iron and nickel into the core when Earth's core formed, which is consistent with the known chemical behavior of lead. This is why McCulloch called the results of his calculations using ancient lead ores the age of Earth's core, rather than the age of Earth.

But it need not be left there. There are other ways of using isotopes of lead and other elements to verify and refine the timing of events in the early Solar System, including the time it took for Earth to form. Those methods are the subject of the next chapter.

CHAPTER NINE

The Formation History of Earth and Meteorites: Sorting Out the Details

Over the past decade, scientists have had some success in determining the sequence and timing of events in the early Solar System using several relatively new isotope tools. These tools are fundamentally of two types. One is the use of precise lead isotope measurements on particular minerals from meteorites. Another involves measuring the distributions of short-lived and now extinct radioactive nuclides, that is, those with half-lives of only a few million years, and their stable daughter isotopes. How these methods work and the history emerging from these efforts are the subjects of this chapter.

Lead Isotopes

In addition to measuring the ages of ancient rocks and minerals, the Pb-Pb method can also distinguish between the ages of ancient events separated by only a few million years. This latter capability works for several reasons. One is that the method utilizes two isotopes of the same element, thereby avoiding potential errors caused by the natural chemical fraction of different elements and by having to measure isotope amounts in the laboratory. Another reason is that the half-life of ^{235}U is relatively short (704 million years), so the ratio of its stable daughter ^{207}Pb to ^{206}Pb, the daughter of ^{238}U, was changing rapidly 4.5 billion years ago when ^{235}U was abundant. Finally, modern methods for eliminating lead contamination and for measuring lead isotopes are extremely good. Applying the method to minerals that are very high in uranium and low in lead can further enhance these advantages. In such minerals, the amount of initial lead is small or negligible, and nearly all the lead in the minerals is the result of the decay of uranium. The use of such minerals also means that any errors in the assumed composition of the initial lead have essentially no effect on the calculated ages.

A good example of the use of lead isotopes to measure the ages of early events in the Solar System is the recent research of Claude Allègre, Gérard Manhès, and Christa Göpel of the Laboratory of Geochemistry and Cosmochemistry at the University of Paris. They have analyzed small meteorite fragments and selected minerals from meteorites that are relatively high in radiogenic lead and low in initial lead, as indicated by having $^{206}Pb/^{204}Pb$ ratios exceeding 150. Their results, which also include a few data from other laboratories, are shown in Figure 9.1.

The carbonaceous chondrite Allende contains a number of small inclusions of unusual high-temperature minerals. Chemical considerations show that these inclusions, called *calcium-aluminum inclusions* (CAIs), must have condensed directly from the Solar Nebula and thus are the oldest objects to have originated in the Solar System. The four inclusions analyzed by Allègre and his colleagues range in age from only 4.565 to 4.568 Ga, with an average age of 4.566 ± 0.002 Ga. This highly precise age, being for the oldest dated event in the Solar System, provides a time signpost—the age to which other ages for early Solar System events can be compared.

The Pb-Pb ages for phosphate minerals from fifteen ordinary chondrites range from 4.504 to 4.563 Ga (see Figure 9.1), but these meteorites have been brecciated and metamorphosed to varying degrees, and most of the ages may have been affected by one or more later impact events. The oldest age, 4.563 ± 0.001 Ga, then, represents the minimum age for the formation of the H (high iron and nickel) ordinary chondrite parent bodies, whereas the lowest age of 4.504 Ga probably indicates the maximum age of the end of the impact events that disturbed the ordinary chondrites. Of the six basaltic achondrites analyzed, three are undisturbed fragments of lava flows, and their ages, which range only from 4.558 to 4.556 Ga, represent the times that lava flows erupted onto the surfaces of their parent bodies. The remaining three achondrites are highly brecciated, and their ages are thought to represent not eruption and crystallization, but the effects of impacts.

The conclusions reached from these precise Pb-Pb data on meteorites are quite simple. The first objects to form from the gas of the pre–Solar Nebula were the CAIs, which crystallized at 4.566 Ga. Within 3 million years or so after the formation of the CAIs, the parent bodies of the chondrites formed.

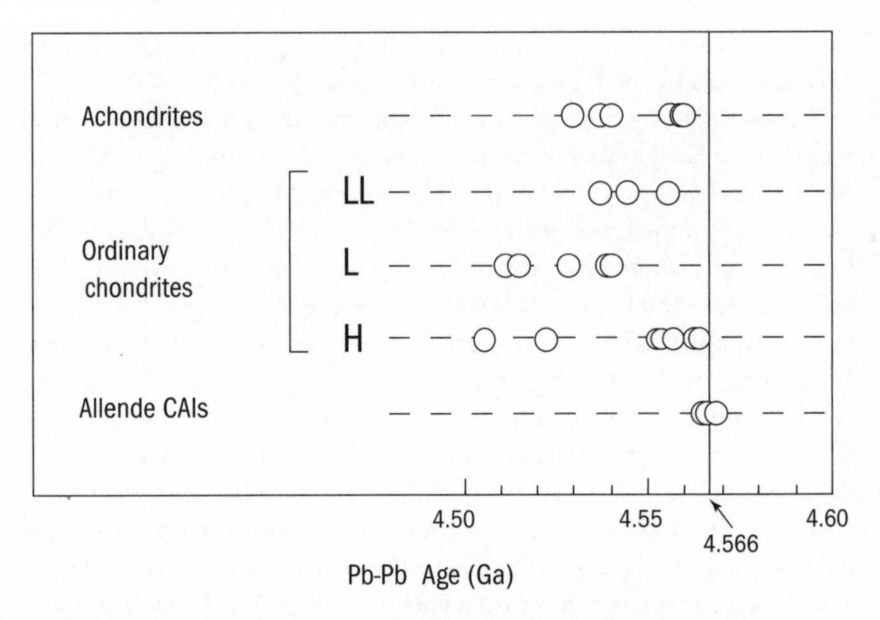

Figure 9.1 A Pb-Pb chronology of high uranium minerals from meteorites with low primordial lead. The calcium-aluminum inclusions (CAIs), dated at 4.566 ± 0.002 Ga and found in some primitive carbonaceous chondrites, are thought to be the first objects to have formed in the Solar System. (After C. Allègre et al., *Geochimica et Cosmochimica Acta*, vol. 59, pp. 1445–1456, © 1995, with permission from Elsevier Science.)

This was followed by the eruption of lava flows onto the surfaces of some of the larger meteorite parent bodies a scant 5 million or fewer years later. The internal melting of a parent body to form lava flows only a few million years after it formed is not as fantastic as it may seem at first. In addition to the gravitational energy collected by parent bodies during their formation, radioactive elements with short half-lives were plentiful in the early Solar System. As they decayed rapidly, they generated enormous amounts of heat, thus allowing the larger bodies to undergo partial melting in their interiors very soon after they formed.

Short-Lived Radioactive Nuclides

The Solar System formed from elements newly created by nearby supernovae. As a result, there was an abundance of radioactive nuclides with half-lives less than 100 million years. These nuclides do not exist today because they have decayed away. Evidence for the existence of almost a dozen of these "extinct" radioactive nuclides exists in the form of anomalous amounts of the stable daughter nuclides they produced. Some of these daughter nuclides can be used to measure small time differences between early Solar System events, but they work a little differently than the radiometric dating methods used to date rocks.

The extinct nuclides, like their long-lived cousins, decay to stable daughter nuclides of another element. For purposes of illustration, consider the short-lived radioactive isotope ^{26}Al (aluminum), which decays to stable ^{26}Mg (magnesium). If aluminum was incorporated into a meteorite before the ^{26}Al had all decayed, which for all practical purposes takes five to ten half-lives, there would be no ^{26}Al in the meteorite, but there would be ^{26}Mg. It would be apparent that the ^{26}Mg was formed from ^{26}Al because it would be present in amounts that were excessive relative to the other isotopes of magnesium. If a meteorite contains no excess ^{26}Mg, it would have to have formed after all the ^{26}Al in the Solar System had decayed. This information could be used to calculate a minimum age difference between the time of origin of the two meteorites. Another meteorite that contains some excess of ^{26}Mg, but not as much as in the first meteorite, would have to have formed after the first meteorite and before the second. Therefore, differences in isotope composition indicate differences in time relative to the start of chemical fractionation in the Solar System. A few examples of the use of extinct radioactivities and the conclusions they lead to will no doubt be helpful.

ALUMINUM AND MAGNESIUM

The aluminum isotope ^{26}Al decays to the magnesium isotope ^{26}Mg with a half-life of only 730,000 years, so any of this aluminum isotope that was in existence when the Solar System formed has long since disappeared. A group of researchers, led by Gerald J. (Jerry) Wasserburg and consisting of scientists from the California Institute of Technology and the Smithsonian Institution, has looked for evidence of extinct ^{26}Al in meteorites. In order to maximize the sensitivity of their experiments, they selected mineral phases from chondrite meteorites that were high in aluminum relative to magnesium. They found that CAIs in carbonaceous and ordinary chondrites contained substantial excess ^{26}Mg from the decay of ^{26}Al. They also discovered that some chondrules from ordinary chondrites contain a small amount of excess ^{26}Mg, but most do not. From these data they concluded that the chondrules containing excess ^{26}Mg (and therefore originally incorporated some ^{26}Al but not as much as the CAIs) must have formed about 2 million years after the CAIs, and that the ^{26}Mg-free chondrules formed at least 1–3 million years later.

These conclusions require that the differences in ^{26}Al content in the CAIs and the chondrules are due to differences in their formation times, not to inhomogeneities in the aluminum isotope composition of the Solar Nebula, but there are data and arguments to support this supposition. The data also show that the time between the creation of the ^{26}Al in a nearby supernova and its incorporation into meteorites was no more than a few million years, and that the lifetime of the Solar Nebula was only about 5 million years or so.

HAFNIUM AND TUNGSTEN

The half-life of ^{182}Hf (hafnium), which decays to ^{182}W (tungsten), is 9 million years. This isotope pair can be used to determine the time of formation of the iron-nickel cores of planetary bodies relative to the time of formation of carbonaceous chondrites. It works like this. Hafnium and tungsten are equally difficult to melt or vaporize, so bodies that formed early in the history of the Solar System, such as planets and asteroids, should have the same proportion of these two elements as carbonaceous chondrites. Hafnium, however, has a strong chemical affinity for silicate minerals, whereas tungsten likes to follow metals like iron and nickel. As a result, the formation of an iron-nickel core within a planet or asteroid produces a huge fractionation, or separation, of these two elements. If the fractionation occurs before all the ^{182}Hf has decayed, the silicate part of the body, which will have a high proportion of hafnium relative to tungsten, will end up with an excess of ^{182}W relative to carbonaceous chondrites. The metal core will end up with a deficiency of ^{182}W relative to carbonaceous chondrites. For such bodies, it is even possible to estimate the differences in their times of core formation from the relative amount of ^{182}W and estimates of the bodies' hafnium/tungsten ratios. If a planetary body develops a core after all detectable ^{182}Hf has decayed—about 50 million years at the current levels of measurement—then both the silicate part of the body and the metal core will have the same tungsten isotope composition as carbonaceous chondrites.

Alex N. Halliday and his colleagues at the Swiss Federal Institute of Tech-

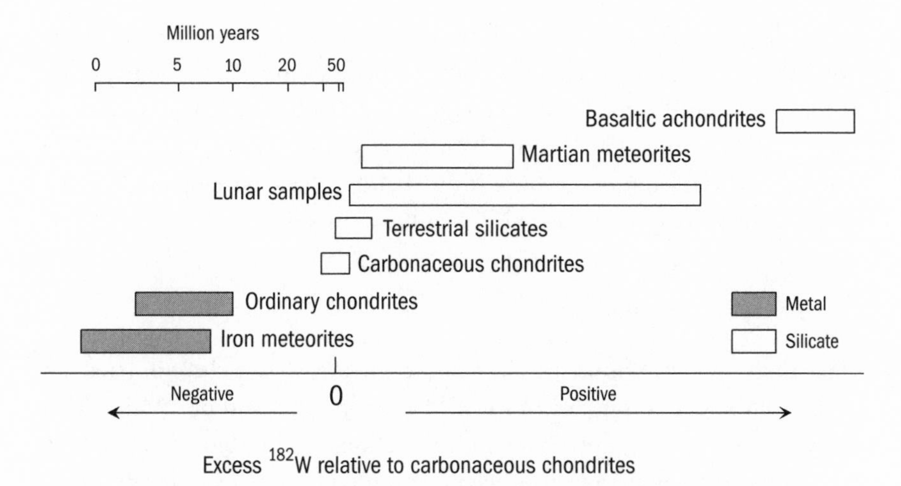

Figure 9.2 Tungsten isotope compositions for various samples of the Solar System. The tungsten isotope ^{182}W is formed by the decay of ^{182}Hf, which is extinct. The upper time scale shows the time differences between the separation of metal from a reservoir of primitive chondrite composition. (Data from Halliday and Lee 1999.)

nology in Zurich have pioneered the development and use of the hafnium-tungsten method. Their results are extremely interesting (Figure 9.2). The first striking thing about these data is that silicates from Earth have, within analytical error, the same proportion of ^{182}W as carbonaceous chondrites, despite the observation that the ratio of hafnium to tungsten in the silicate Earth is about fifteen times that in carbonaceous chondrites. The only reasonable explanation for this is that the formation of Earth's core must have occurred after all the radioactive hafnium had disappeared. This would have been 50 million years or more after the elements in the Solar System formed. In contrast, iron meteorites and metals in ordinary chondrites have deficiencies in ^{182}W, whereas basaltic achondrites have large excesses. Such anomalies are expected if the parent bodies of the iron meteorites, the ordinary chondrites, and the basaltic achondrites formed cores only a few million years after the Solar System began.

Some silicates from the Moon and Mars appear to have tungsten isotope compositions resembling carbonaceous chondrites, whereas others contain excess ^{182}W. Calculations based on these data and estimates of the hafnium/tungsten ratios of these two bodies indicate that the Moon formed about 50

± 10 million years after the start of the Solar System, or about the same time that Earth developed a core. This timing provides support for the hypothesis that the Moon formed from material generated by the impact of a giant asteroid with a proto-Earth about half its present size. In contrast, the data suggest that Mars appears to have formed and to have developed a core by the time the solar system was only 15 million years old.

MANGANESE AND CHROMIUM

Manganese and chromium are abundant elements, so this chronometer can be used on a wide variety of objects. The manganese isotope ^{53}Mn, which decays to ^{53}Cr (chromium), has a half-life of only 3.7 million years, giving the method the power to resolve events separated in time by only a few million years. G.W. Lugmair of the Scripps Institution of Oceanography in California and A. Shukolyukov of the Max Planck Institute for Chemistry in Mainz, Germany, have applied this method to a number of achondrite meteorites and a stony iron meteorite. By comparing the chromium isotope ratios in these samples, they found that some of the meteorites formed while ^{53}Mn was still in existence, while others formed after it had become extinct. This means that the formation of the parent bodies of these meteorites occurred over a period of at least 20 million years.

Another interesting finding of this research is that the parent body of a chemically related family of basaltic achondrites formed, differentiated, and erupted basaltic lava flows onto its surface within a period of less than 3 million years. By calculating the time difference between this parent body and two meteorites that have been precisely dated by Pb-Pb methods, Lugmair and Shukolyukov concluded that the parent body formed almost immediately after the CAIs formed. Based on its surface composition, this parent body is thought to be the asteroid Vesta, the third largest asteroid in the asteroid belt, with a diameter of 550 km.

PALLADIUM AND SILVER

The palladium isotope ^{107}Pd decays to ^{107}Ag (silver) with a half-life of 6.5 million years. Because palladium and silver are both metals, this chronometer is applied primarily to iron and stony iron meteorites. Jerry Wasserburg and his colleagues have investigated this extinct radioactivity. They have discovered that of the nearly three dozen meteorites analyzed by them, all but two appear to have formed within 12 million years of each other. This indicates that the parent bodies of these iron and stony iron meteorites formed and differentiated within a relatively short period of time. Note that the period of time for the formation of differentiated bodies indicated by this method is only about half that indicated by the Mn-Cr isotope pair, a matter to be resolved by further research.

A Tentative Chronology of the Early Solar System

The detailed chronology provided by lead isotopes and by extinct radioactivities is an area of vigorous research, and we still have much to learn. Nevertheless, a tentative if imperfect chronology of the early Solar System is beginning to emerge (Figure 9.3). Some of the uncertainties in this history are due to apparent conflicts between the different isotope chronometers. For example, the lead isotope data suggest that the basaltic achondrite parent bodies began to form about 8 million years after the CAIs, whereas the data from the Mn-Cr system suggest that at least one basaltic achondrite parent body, perhaps the asteroid Vesta, formed almost immediately after the CAIs. Such apparent discrepancies are most likely due to incomplete data and to the possibility that the events measured by the different isotope systems are imperfectly understood.

The earliest datable event in the Solar System is the formation of CAIs, which marks the beginning of condensation of solid material from the Solar Nebula. The CAIs found in the carbonaceous chondrite Allende have been precisely dated by the Pb-Pb method as 4.566 Ga in age.

It seems likely that the principal bodies in the Solar System began to aggregate and grow within a very short period of time after solid matter condensed from the Solar Nebula. Indeed, theory indicates that planetary embryos would accumulate in only 100,000 years if solid material was available. Exactly when this aggregation began is uncertain, but it was likely less than a few million years after the formation of CAIs. Thus, it seems probable that the parent bodies of the chondrites, the achondrites, the stony iron meteorites, and the iron meteorites began to form at nearly the same time. It also seems that the parent bodies of the latter three rapidly grew large enough to melt partially or wholly, to differentiate, and to form iron-nickel cores. This process continued for at least 12–15 million years, and

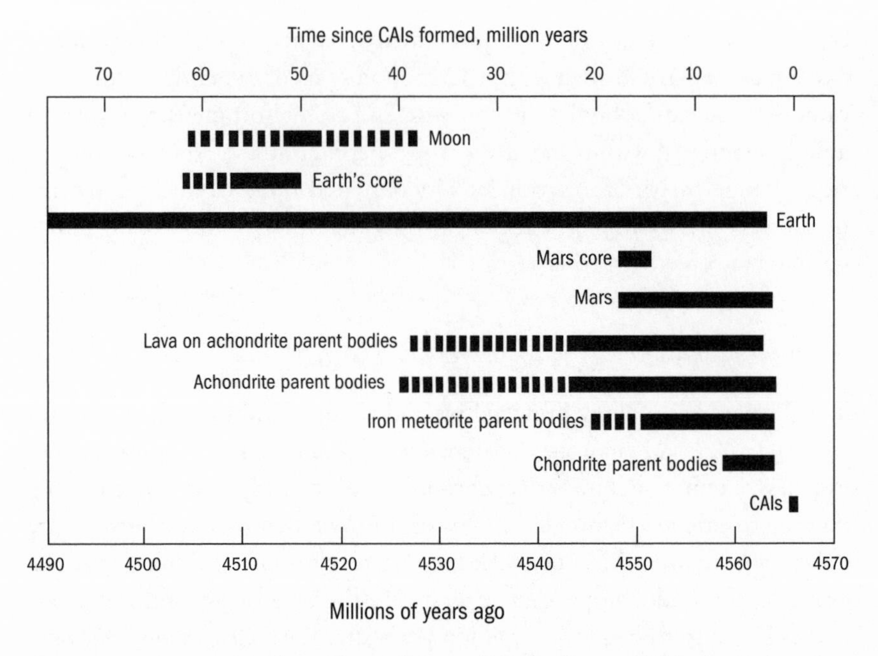

Figure 9.3 A tentative chronology of the early Solar System. This chronology is based on isotope data from lead and from extinct radioactive nuclides.

perhaps longer. Isotope data on meteorites from Mars indicate that Mars also began to aggregate very early, formed its core within 15 million years or so, and then ceased growing.

There are no isotope data to indicate when Earth began to grow, but it seems reasonable to think that its embryo formed near its present orbit about the same time as the other major bodies, only a few million years or less after formation of the CAIs. The growth of Earth, however, occurred over a long period of time. The hafnium-tungsten data clearly show that Earth's core did not form until about 50 million years after the CAIs, most likely at about the same time as, and perhaps aided by, the collision with Earth of the Mars-sized body that led to the formation of the Moon. Earth, and probably the Moon, then continued to grow for several tens of millions of years more, until Earth reached its present size at about 4.49 Ga, the age of Earth's core indicated by the most ancient lead ores and the reconstruction of the terrestrial lead growth curve. One consequence of this relatively long time scale for the growth of the planet is that there is no longer a single number that adequately represents the age of Earth. Instead, scientists must now think in terms of the ages of specific events in Earth's formation history. This makes discussion of the "age of Earth" somewhat more complicated than it was a decade ago—but also more informative, interesting, and satisfying.

No doubt the sequence and timing of events in the early Solar System will be greatly refined as additional isotope measurements are made and their meanings become better understood. Nevertheless, because of new techniques and improved instrumentation, scientists know much, much more about the formation of the planets and asteroids than they did only a decade ago.

CHAPTER TEN

The Ages of the Universe, the Galaxy, and the Elements

The preceding chapters summarized the evidence from radiometric dating for the ages of Earth, the Moon, and the meteorites. The results lead to the inescapable conclusion that the solid bodies of the Solar System began to form 4.566 billion years ago and that the principal rocky bodies of the Solar System were fully assembled within about 50 million years thereafter. As convincing as the evidence is, however, scientists like to have independent checks on important data and conclusions whenever possible. So it is reasonable to ask whether there is confirmatory evidence for the age of Earth from methods that do not rely on radiometric dating. There is. The evidence is found in the properties of the very small—the elements; and of the very large—the stars and galaxies.

It is axiomatic that the age of the Galaxy must be equal to or greater than the objects it contains, and the same reasoning may be applied to the Universe. There are several ways to estimate the ages of the Milky Way Galaxy, the elements, and the Universe. There is also a way to calculate the age of the Sun that does not depend on radioactivity. These methods provide further evidence that our cosmic surroundings are billions of years old. It is worthwhile to review the more important methods and the insights they provide about the age of Earth and the Solar System.

The Expansion of the Universe

As discussed in Chapter 2, the Universe is expanding. A result of this expansion is that each of the galaxies in the Universe is moving away from all the other galaxies. As observed from Earth, therefore, all of the distant galaxies are receding. An observer in any other galaxy would see the expansion in exactly the same way as we do.

Scientists can measure the velocity of distant objects as they move either toward or away from Earth using the *Doppler effect*. As any source of light moves toward or away from an observer, there is a velocity-dependent shift in both the wavelength and the frequency of the light. For objects moving toward the observer, the wavelength is shortened and the frequency is increased. For objects moving away, the wavelength is lengthened and the frequency is decreased. Wavelength defines color, so for a moving source of light, the color of the light is shifted toward the blue end of the spectrum for an approaching source (a *blue shift*) and toward the red end of the spectrum for a receding source (a *red shift*). This phenomenon occurs for all types of radiation, including sound. No doubt you have noticed the change in the pitch of a train whistle from higher to lower as the train approaches (higher pitch) and then recedes into the distance (lower pitch).

In 1912, V.M. Slipher of the Lowell Observatory near Flagstaff, Arizona, observed a red shift in the light from galaxies and interpreted it as relative motion of the galaxies away from the Milky Way Galaxy. Seventeen years later, Edwin P. Hubble of the Mt. Wilson Observatory in California made the important observation that the motion of galaxies away from the Milky Way was the same in all directions. This movement of the galaxies away from each other is called the *Hubble flow*. Hubble also found that the relative velocity of a galaxy was proportional to its distance multiplied by a constant—in other words, the farther away the galaxy, the faster it is receding. This relationship has been repeatedly verified and is known as *Hubble's law* and the constant, H_0, as the *Hubble constant*. The Hubble constant is found from the slope of the straight line that results when the velocities of galaxies are plotted as a function of their distance (Figure 10.1).

Hubble's law describes an interesting relationship. Not only does it relate velocity to distance, but it also can be used to calculate the age of the Universe. This is because the Hubble constant is in units of 1/time (velocity divided by distance), so $1/H_0$, which is known as the *Hubble time*, is the time since the expansion of the Universe began, provided that the expansion has neither speeded up nor slowed down. This assumption that the expansion has been linear is now thought to be incorrect, but we'll discuss that point later.

As it turns out, the units of the Hubble constant are rather convenient—

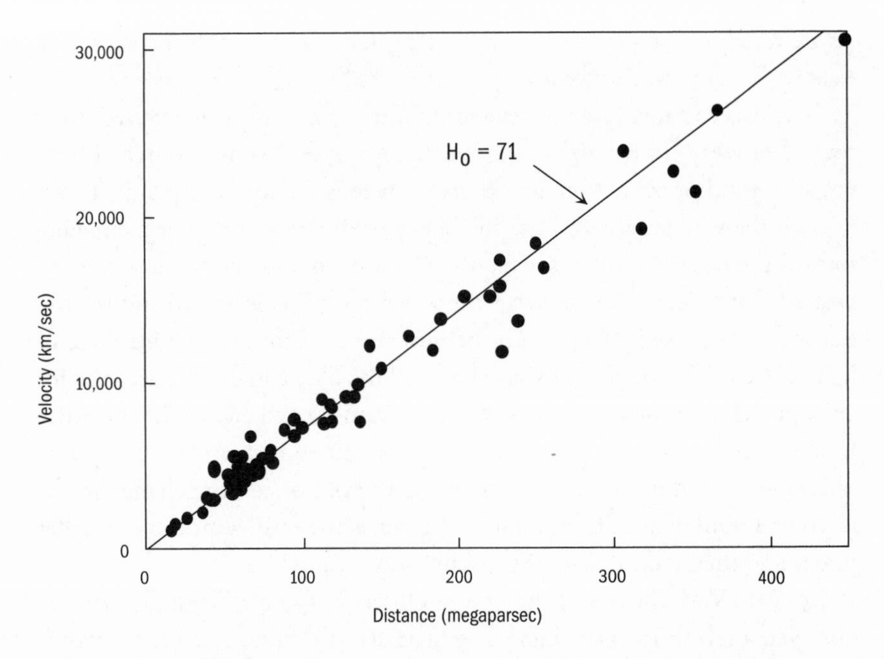

Figure 10.1 A Hubble diagram for distant galaxies, measured using type Ia supernovae. The slope of the line (velocity/distance) gives the Hubble constant, H_0, which has units of 1/time and thus is a measure of the age of the expanding Universe. Each distance was determined by one of four different methods. (After W.H. Freedman, *Physics Reports*, vol. 333-334, pp. 13–31, © 2000, with permission from Elsevier Science.)

when distance is expressed in kilometers and time in years, the result is that the Hubble time multiplied by 1000, $1000/H_0$, is then in units of billion years to within a few percent. The errors in the estimates of H_0 are much larger than a few percent, so it is customary to use $1000/H_0$ as the Hubble time in billion years. For the data in Figure 10.1, the Hubble time is 1000/71 = 14.1 Ga. If it were only that simple!

Although Hubble's law is straightforward and uncontroversial, its application to the Universe is more complicated. The measurement of relative velocities using the Doppler effect is not difficult, but superimposed on the velocities purely due to expansion of the Universe are the local motions of stars within galaxies, of galaxies within groups of galaxies, and of groups of galaxies relative to each other. For very close objects, these local motions are large compared to the Hubble flow. For more distant objects, from a few to a few tens of megaparsecs away, corrections can be made for the local motions. (A megaparsec is a unit of distance equal to 3.1 x 10^{19} km, or 3.26 million light years.) For objects with distances that exceed 50 megaparsecs, the local motions become negligible, so the most accurate velocities are obtained for objects at considerable distance from the Milky Way.

The distances to faraway objects, such as other galaxies, are more difficult to measure than their velocities, so it is the distances that contribute most of the uncertainty to the Hubble constant. There are a number ways to measure distances in the Universe, and they are of varying usefulness and range. Most are based on the concept that certain types of objects, regardless of their location, give off the same amount of radiant energy (light) and therefore have the same brightness, or luminosity. Thus, it is possible to calculate distances to some types of objects using their apparent brightness, because the intensity of the light received on Earth diminishes with distance in a precisely known way.

Some kinds of individual stars can be used as distance indicators. One of the most widely used is the Cepheids, a class of variable or pulsating stars whose average luminosities are a function of their rate of pulsation. Individual stars, however, are useful only at distances of a few megaparsecs. Beyond that, individual stars cannot be seen, and scientists must rely on the properties of more luminous objects, such as supernovae and galaxies. Despite their limited range, the Cephids are valuable in calibrating other distance indicators that can be used over greater distances.

One of the more valuable distance indicators is a kind of supernova known as type Ia. Supernovae are created when some kinds of stars reach a final stage in their evolution and explode. These explosions create an almost instantaneous flare of intense light that then decays away within a few weeks. The light from a supernova is so bright that these objects can be seen over enormous distances. All type Ia supernovae have nearly the same luminosity. What differences there are can be correlated with their decay periods and corrected, so they are one of the more accurate of the astronomical distance indicators.

Still another way to measure distance uses the effect of gravity. If the light from a distant galaxy passes close enough to another galaxy or group of galaxies, the large gravitational field may act as a lens. The result is sometimes a double image of the galaxy being viewed. Because the light of each image follows a different path, there is a time delay between the receipt of the images on Earth. Since the speed of light is accurately known, the time delay provides a measure of the distance to the galaxy whose image has been doubled, provided that the gravitational effect can be modeled accurately.

Over the past seven decades, estimates of the Hubble constant have ranged widely. Within the past few years, however, techniques, theory, and knowledge have greatly improved, and values of the Hubble constant obtained since 1998 have been in the range of 60–71 with the more recent being about 71 ± 7. The result is a Hubble time of 14.1 billion years. Although this is not exactly the age of the Universe, it is close.

The Hubble time would be the age of the Universe if the rate of expansion has been constant since the Big Bang. But that could be true only if there were no mass, and hence no gravity, in the Universe to slow down the expansion over time. The very existence of stars and galaxies, however, means that there is mass in the Universe, and the gravitational attraction caused by this mass must have had some slowing effect on the expansion, at least early on. But how much mass is there, and how much slowing has it caused? One hypothesis, which was proposed jointly by Albert Einstein and the Dutch astronomer Willem de Sitter in 1931, is that the mass of the Universe is just enough to eventually bring the expansion to a halt. If this were so, then the age of the Universe would be two-thirds of the Hubble time. This would mean that a Hubble constant of 71 would indicate an age of only about 9 billion years. Recent observations and calculations, however, indicate that the amount of matter in the Universe is only one-fifth to one-third of the mass necessary for an Einstein–de Sitter Universe. Thus, the Universe must be older than 9 billion years.

Another quantity has an effect on the calculation of the age of the Universe from the Hubble constant. This quantity is the *vacuum energy* (also called dark energy) of the Universe, a force that repels objects and therefore counteracts gravity. The vacuum energy is a direct consequence of the cosmological constant, a factor introduced in 1917 by Einstein into his equations for general relativity in order to explain why gravity does not cause the Universe, then thought to be static, to collapse. Once expansion was discovered and confirmed, this was no longer a problem, and Einstein thought his introduction of a cosmological constant probably had been a mistake. But today it seems that he may have been right in the first place. Recent observations of some very distant supernovae have shown that there is a pervasive force in the Universe that counteracts gravity, and Einstein's cosmological constant may, after all, be necessary.

Vacuum energy has an interesting effect on expansion. It may seem bizarre, but the density of vacuum energy does not diminish as the Universe expands.

Its effect is always constant. This is in contrast to matter, which becomes more dilute (less dense on average) as expansion proceeds. As matter becomes more and more dilute, the overall effect of gravity on the Universe decreases. Initially, when the matter was close together, the expansion was continuously slowed by the effect of gravity. As the galaxies moved apart, however, the effect of gravity decreased so the effect of vacuum energy, which is constant, became relatively more and more important over time. Eventually, vacuum energy dominated gravity, and the rate of expansion began to accelerate. Recent evidence indicates that the Universe has followed just such a scenario.

In 1998, two groups of astronomers, the High-Z Supernova Search Team and the Supernova Cosmology Project, reported the first evidence that the rate of expansion of the Universe has not been constant. They observed that several dozen type Ia supernovae appeared dimmer, or farther away, than their velocities indicate if the expansion of the Universe was either constant or decelerating. Their conclusion was that the Universe has been accelerating for the past 6 billion years. It was this observation that convinced scientists of the existence of vacuum energy, or something that acts very much like it.

In 2001, the High-Z Supernova Search Team published another remarkable conclusion from observations of the most distant type Ia supernova ever seen, called SN1997ff. This supernova, seen in 1997, occurred when the Universe was only about one-third of its present size and one-fourth of its present age. Because the Universe is so vast and the speed of light is finite, astronomers who view distant objects are also looking back in time observing events, like SN1997ff, that occurred billions of years ago. Based on its velocity, SN1997ff appears much brighter, or closer to us, than it would if the rate of expansion had been constant. From this the team concluded that the early Universe was indeed decelerating, just as the existence of relatively low total mass in the Universe and significant vacuum energy require.

The early deceleration and subsequent acceleration of the expansion complicates calculations of the age of the Universe based on the Hubble constant. Over the past several years, however, scientists have finally been able to make good estimates of the mass of the Universe and of the magnitude of the vacuum energy. The result has been decreasing uncertainty about the age of the Universe. Estimates of the age of the Universe from Hubble expansion made since 1998 fall in the relatively narrow range of 13–15 billion years, with uncertainties of 2 billion years or less. Only a decade ago estimates as low as 7 billion years and as high as 20 billion years could not be ruled out, so this is quite an improvement. There is also little doubt that additional astronomical observations will lead to better estimates of the age of the Universe in the very near future. For now, however, it appears that the Universe is about three times older than the Solar System and Earth.

An Echo of the Big Bang

As mentioned in Chapter 2, Arno Penzias and Robert Wilson of Bell Laboratories discovered the cosmic microwave background in 1965 and were rewarded with the Nobel Prize in Physics. This microwave radiation originated in the Big Bang, and it is a measure of the average temperature of the Universe, which has been cooling as the Universe has expanded. This phenomenon is similar to the cooling of compressed gas as it is released from a pressurized container. The cosmic microwave background now indicates a temperature of only 2.7 K. Sound waves in the forming Universe left their imprint on the microwave radiation in the form of slightly warmer and slightly cooler spots, which correspond to regions where the radiation is being compressed and expanded. The size of these spots is uniform throughout the Universe. The older the Universe, the farther away these spots are and the smaller they appear.

In 2001, Lloyd Knox and Constantinos Skordis of the University of California at Davis, and Nelson Christensen of Carlton College, analyzed the size of the cooler and warmer spots in the cosmic microwave background and calculated that the age of the Universe is 14.0± 0.5 billion years. This age determination requires no knowledge of either the Hubble constant or the cosmological constant, but it does require that the Universe is "flat" so that the size measurements are accurate. In a flat Universe the laws of Euclidean geometry apply—parallel lines never meet and the relationships between angles are the same as taught in geometry class. Other measurements on the cosmic microwave background and on type Ia supernovae confirm that the Universe is flat or very nearly so, so the age of the Universe found from the cosmic microwave background is probably not seriously in error.

The Oldest Stars in the Milky Way

The Milky Way is a spiral galaxy composed of billions of stars (see Figure 2.1). Like a giant pinwheel, it rotates about its brilliant central bulge once every few hundred million years. But not all of the stars are confined to the nucleus and the flattened disk. Scattered around a spherical region surrounding the Milky Way are billions of additional faint stars. Within this "halo" are globular clusters, of which there are some 200, each containing a hundred thousand or more stars. The stars in globular clusters have very low fractions of their mass in elements heavier than helium. This is because the globular clusters formed early in the history of the Milky Way, before the Galaxy had collapsed into a flattened disk, and before recycling of elements through high-mass stars and supernovae had substantially increased the heavy element content of the material from which stars are made. For this reason, the stars in the globular clusters are thought to be the oldest stars in the Milky Way Galaxy, and there is a way to tell how old they are.

Stars are born when gravitational forces within a cloud of gas and dust cause the cloud to collapse. The gravitational energy released by the collapse generates tremendous internal temperatures that eventually cause the hydrogen in the core of the newborn star to "ignite" and "burn." The burning that occurs in the interior of a new star, however, is not ordinary combustion, but rather nuclear fusion reactions that convert the hydrogen into helium. These reactions release enormous amounts of heat and light.

Stars come in a variety of sizes, and the Sun is average in size. Stars range from about 0.08 to as much as 60 times the mass of the Sun. Two of the observable properties of stars are their luminosity (brightness), usually measured relative to the luminosity of the Sun, and effective temperature, which is the surface temperature the star would have if it were a perfect radiator of energy. The luminosity of a star can be determined from its apparent brightness and distance, and the effective temperature from its color.

When luminosity is plotted as a function of effective temperature, most stars fall within a narrow band called the *main sequence* (Figure 10.2). This relationship was first discovered in 1911 by the Danish astronomer Ejnar Hertzsprung and independently in 1913 by Henry N. Russell of Princeton University (the same scientist who first calculated an age for Earth's crust; see Chapter 3). This type of graph is therefore called a *Hertzsprung-Russell (H-R) diagram.* Stars are born and spend most of their lifetime on the main sequence. The single most important property that determines where any newborn star will fall on the main sequence and how long it will remain there is its mass. Both theory and observation indicate that the greater the mass of a main-sequence star, the brighter, hotter, and shorter-lived it will

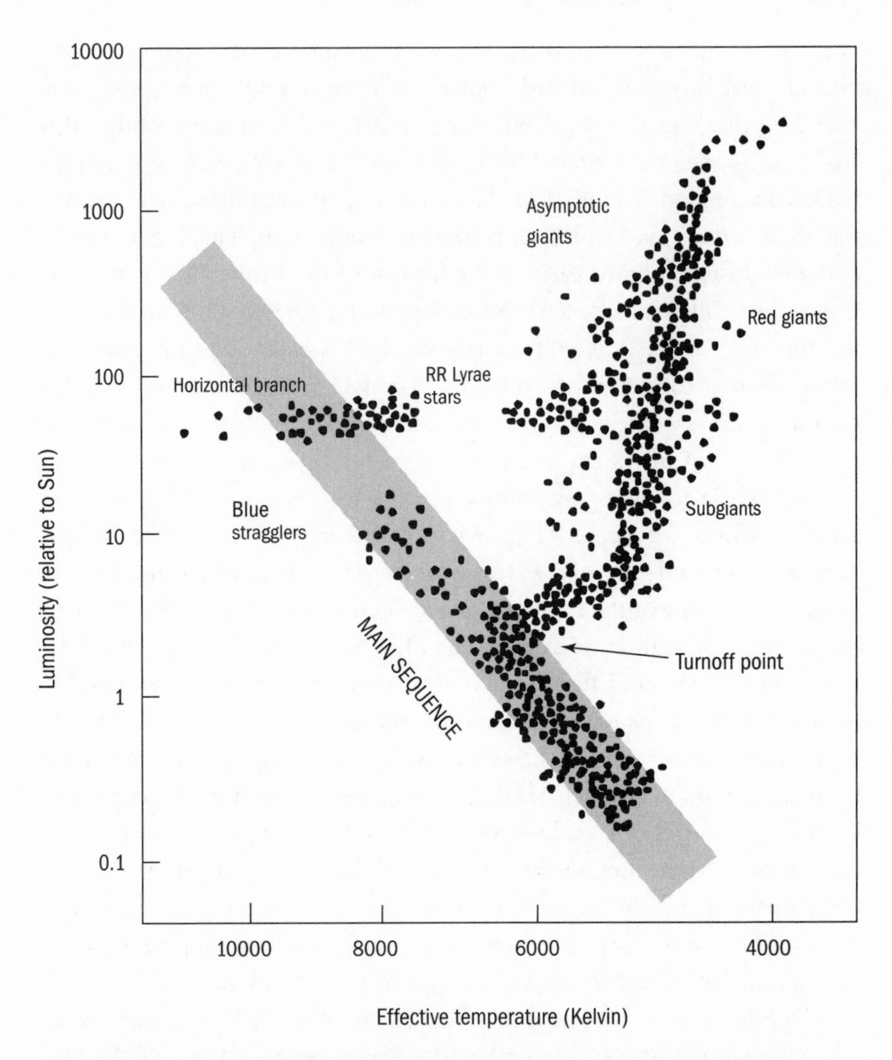

Figure 10.2 A Hertzsprung-Russell diagram for stars in the globular cluster M3. Luminosity is measured relative to the Sun, while the effective temperature is a measure of a star's surface temperature as determined from its color. Stars spend most of their lifetime on the shaded band known as the main sequence. Scientists can calculate the age of a globular cluster from the position of the turnoff point, which progresses downward along the main-sequence band over time. (After Shu 1982; based on data from Johnson and Sandage 1956.)

be. Thus, the highest mass stars plot at the upper (left) end of the main sequence and the lowest mass stars at the lower (right) end.

Once a main-sequence star has exhausted its hydrogen fuel, it leaves the main sequence and rapidly evolves through a series of changes that ultimately lead to its death. These changes are a result of the exhaustion of successive rounds of nuclear fuel, and the delicate balance between the enormous pressure of gravity, which works to collapse the star, and the increasing resistance to compression as the star heats up from the pressure of gravity. The precise evolutionary path a star follows depends on its mass and, to a lesser degree, the details of its composition. Both theory and observation indicate that stars with a mass less than about six times that of the Sun all follow similar evolutionary paths, leaving the main sequence to become red giants. This will be the fate of the Sun in another 5 billion years or so. High-mass stars, on the other hand, evolve more rapidly along quite different tracks in the H-R diagram, becoming supernovae in less than 10 million years. Low-mass stars, which evolve much more slowly than high-mass stars, are used to determine the ages of globular clusters.

Based on the physics of the various nuclear reactions, energy transfer, and gravitational effects that occur in stars of a given mass and composition, scientists can calculate how long a star will remain on the main sequence. They can also determine its evolutionary track once it leaves the main sequence and the rate at which it progresses along that track. These calculations provide the basis for estimating the ages of globular clusters.

All the stars in a globular cluster are about the same age, so their distribution on an H-R diagram is due primarily to differences in their mass, not to differences in their age. Since the more massive stars have a shorter lifetime, they leave the main sequence before the less massive stars. Thus, over time stars progressively "peel off" of the main sequence to the right, much like the peel is pulled from a banana. This results in a "turnoff point" on the H-R diagram of a globular cluster (see Figure 10.2). The position of the turnoff point is very sensitive to age, and it is from the position of this turnoff point that the age of a globular cluster can be found.

Calculations since 1997 of the age of the Milky Way Galaxy using data from globular clusters fall in the range of 11.5–14.0 billion years and typically have estimated errors of 2 billion years or less. Since globular clusters formed very early in its history, the Milky Way must be slightly older than this, but probably not much older.

Cool White Dwarfs

A white dwarf is a very small, dense, hot, and dim former star whose nuclear fires have gone out. White dwarfs have a density about a million times that of water; a teaspoon of their matter would weigh 5 tons. Running on left-over heat, they appear dim only because of their small size. Some 98% of all stars, including the Sun, will end their lives as white dwarfs after running their evolutionary course as stars. Because their temperature cannot be sustained, white dwarfs cool over time. The temperature of a white dwarf can be measured by the amount of light it gives off—the cooler it is, the dimmer it becomes. Like anything that cools, the cooling rate of a white dwarf slows as its temperature decreases, so the Galaxy should contain more numerous cooler white dwarfs than warmer white dwarfs, and it does. There is a sharp decrease, however, in the number of white dwarfs that have very cool temperatures. The only reasonable explanation for this decrease is that white dwarfs have had only a finite time to cool, so the time it took for the few coolest (and oldest) white dwarfs to cool must be a minimum age for the Milky Way Galaxy.

In 2002, a group of scientists, led by Harvey Richter of the University of British Columbia and Brad Hansen of the University of California at Los Angeles, published observations, made with the Hubble Space Telescope, of white dwarfs in the globular cluster M4. These are the faintest and coolest white dwarfs ever observed. The data and calculations of Richter, Hansen, and their colleagues show that the white dwarfs in M4 are 12.7± 0.7 Ga, which is also an approximate age for the Milky Way Galaxy. The age for white dwarfs in M4 is completely independent of the ages of globular cluster stars determined from their turnoff point on the main sequence and in excellent agreement with it.

The Vibrating Sun

The outer 30% of the Sun is in constant motion, like a pot of boiling water, and the turbulence of this motion creates seismic waves that propagate through the interior of the Sun. As these waves reflect off the surface from beneath, they cause surface disturbances that resemble the waves on the surface of the ocean. These surface disturbances can be observed from Earth using Doppler effect imaging instruments. Since the propagation of seismic waves depends on density, which is a function of pressure and composition, the waves observed at different frequencies provide important information about the interior structure of the Sun. They also provide information about the Sun's age, because the interior structure of a star changes as it consumes its nuclear fuel.

Combining observations of the Sun's vibrations with refined solar evolution models, David B. Guenther of St. Mary's University in Halifax and Pierre Demarque of Yale University have calculated an age for the Sun of 4.5± 0.1 Ga. Nuclear reactions in the Sun probably began some tens of millions of years after the earliest meteorites formed, so this age is in excellent agreement with the 4.566 Ga age of the oldest meteorites.

The Age of the Elements

It is possible to estimate the age of the elements using certain long-lived radioactive nuclides. The general approach is to calculate the length of time required to produce the present ratio of a pair of radioactive nuclides, called a chronometer pair. This approach can be used even if only one of the nuclides of the pair is radioactive because the ratio of the pair will still change over time. Three things must be known for the method to yield an age: the ratio of the nuclide pair when the Solar System formed, the rate at which each nuclide is formed by nuclear processes, and the production history of the nuclides. The ratio of a chronometer pair when the Solar System formed can be found from analyses of meteorites, lunar rocks, and terrestrial samples. The production ratios and history, however, are a bit more complex to estimate.

Elements form in stars and in catastrophic astrophysical events by a number of nuclear processes. Nuclear fusion reactions cause specific light elements to form slightly heavier elements within ordinary stars. The nuclear "burning" of hydrogen, helium, carbon, oxygen, and silicon occur more or less successively and at progressively higher temperatures as stars evolve. Each process proceeds to the next after it has exhausted its fuel supply, provided that the mass of the star is sufficient to generate the higher temperatures necessary to sustain the nuclear reactions of the next step in the sequence. These nuclear processes lead to elements up to and including iron, but they are incapable of generating heavier elements.

Most of the nuclides heavier than iron are synthesized when lighter nuclides capture neutrons. The two most important processes by which this occurs are the s-process (slow neutron capture) and the r-process (rapid neutron capture). Both of these processes result in the formation of increasingly neutron-rich nuclei of an element followed by radioactive decay to form other elements. The principal difference between the two processes is the rate of neutron addition and radioactive decay.

The *s-process* occurs when the rate of addition of neutrons to the available nuclei is very slow compared to the lifetimes of the unstable nuclides formed by the neutron additions. As successive neutrons are added, isotopes of an element of progressively higher mass are formed, until a radioactive isotope is created. At this point, the radioactive isotope will decay as quickly as it is formed, so that the isotope of next higher mass cannot form. The s-process can generate a large number of nuclides from the relatively small variety of seed nuclei produced by the burning processes in stars discussed above, but because of the blocking caused by the decay of short-lived radioactive nuclides, not all elements can be formed in this way. The s-process is thought to occur primarily in red giant stars, where there are sufficient free neutrons to drive the s-process, but insufficient neutrons for the r-process.

The *r-process* occurs when neutrons are added more rapidly than the unstable isotopes that are formed can decay, so that blocking by radioactive nuclides cannot occur as it does in the s-process. The rapid addition of neutrons to existing nuclides forms a wide variety of neutron-rich radioactive nuclides. The formation of these nuclides continues to progressively higher masses until the half-lives of the nuclides are exceedingly small, and equilibrium between production and decay is established. Only at this point is the r-process blocked by radioactive decay. Each of the unstable nuclides formed by this rapid addition of neutrons decays in time, forming nuclides with progressively higher numbers of protons and lower numbers of neutrons until the process is blocked by a stable nuclide. The r-process can occur only where the rate of neutron production is exceedingly high. This does not happen in stars, so the r-process seems to require some sort of astrophysical catastrophe. One place where the r-process occurs is in supernovae, but it may also occur in other catastrophic events as well.

Virtually every known nuclide can be produced in greater or lesser amounts by the known nucleosynthethic processes. Although the r-process and s-process are the most effective at manufacturing new elements, there are other processes, and in nature, nucleosynthesis proceeds through the complex interaction of several of the processes operating simultaneously.

Which processes occur at any given time depends on the mass and maturity of the star, and on the material from which the star is originally formed. Stars that become supernovae provide material, including new elements, from which other stars eventually form. Thus, succeeding generations of stars are progressively enriched in the heavier elements. This is the reason that the ancient globular clusters have lower contents of the heavy elements than the younger stars that reside in the galactic plane.

Several chronometer pairs have been used to find the age of the elements in the Milky Way Galaxy, but most involve nuclides that are created by more than one mechanism, which greatly complicates determining their production ratios. The chronometer pair that involves only one creation process is $^{232}Th/^{238}U$. Both of these nuclides are created only by the r-process, and they also have half-lives long enough to be of use in measuring the age of the Galaxy. The ratio of their production rates is known to within about 10%, and their abundance in the Solar System is known from analyses of meteorites, lunar rocks, and terrestrial samples. This leaves only the production history to figure out.

One approach to production history is to assume that uranium and thorium were produced in a single, short event when the Milky Way Galaxy formed. Calculations based on this assumption provide a minimum age for the Milky Way because subsequent production of these elements is not considered. When the Solar System formed, however, a number of short-lived radioactive nuclides created by the r-process were incorporated into the planets and meteorites. These nuclides must have been created only shortly before the formation of the Solar System, which means that element production could not have been a single, early event. Thus, the elements from which the Solar System was made were undoubtedly the result of a prolonged process involving the births and deaths of many stars.

For a variety of very good reasons, it is thought that nucleosynthesis occurs on a continuing basis and at an approximately uniform rate as stars are born, evolve, and die. Recent calculations of the age of the elements based on the present ratio of $^{232}Th/^{238}U$, and a uniform rate of production of these nuclides since the Galaxy formed, yield times of 7.9–8.4 Ga. But because there has been no creation of these nuclides within the Solar System (the Sun does not make them), the age of the Solar System must be added to obtain the age of the Milky Way Galaxy. The result is 12.5–13.0 Ga, with an uncertainty of about 3 billion years.

The Case of the Missing Nuclides

There is another line of reasoning that does not provide a specific number for the age of the elements, but it shows that they must be a few billions, not millions or thousands, of years old. Inspection of any chart or table of the nuclides reveals an interesting fact: Of the radioactive nuclides not currently being produced in the natural environment or by humans, only those with half-lives greater than 82 million years occur in the Solar System.

To illustrate, there are thirty-four radioactive nuclides with half-lives greater than 1 million years. Of these, twenty-three are found in nature. Five of the twenty-three, however, are continually being produced by ongoing natural nuclear reactions. For example, ^{53}Mn (manganese) and ^{10}Be (beryllium) are produced primarily in dust particles in the upper atmosphere and in space by cosmic rays. The seventeen remaining radioactive nuclides all have half-lives of 82 million years or more. The absence of short-lived radioactive nuclides in nature holds true even for nuclides with half-lives shorter than one million years—only those whose existence is due to continual production by natural processes (or in some cases by humans) are found.

Three hypotheses have been proposed to explain this curious absence of short-lived radioactive nuclides. The first is that the observed distribution might be due to chance. The second is that the processes that created the elements could be incapable of producing the short-lived nuclides. The third is that a great length of time may have passed since the elements were created, and nuclides with half-lives less than 82 million years have simply disappeared over time because of radioactive decay.

The first hypothesis is easily dispatched by finding the probability that the observed distribution could occur by chance. This probability is not difficult to calculate; it is about 10^{-21} if the elements are only 10 million years old and 10^{-53} if the elements are only 1000 years old. Thus, the hypothesis of pure chance to explain the absence of the short-lived unstable nuclides is very unlikely. Probability, of course, is incapable of completely disproving the hypothesis—there is still that one chance in 10^{21}—but it is so improbable that it can be excluded from further consideration, particularly because there is a much better explanation.

The hypothesis that the missing nuclides were never created is also easily eliminated. Theoretically, the r- and s-processes should easily synthesize most of the missing nuclides. Even nuclides that are less susceptible to cre-

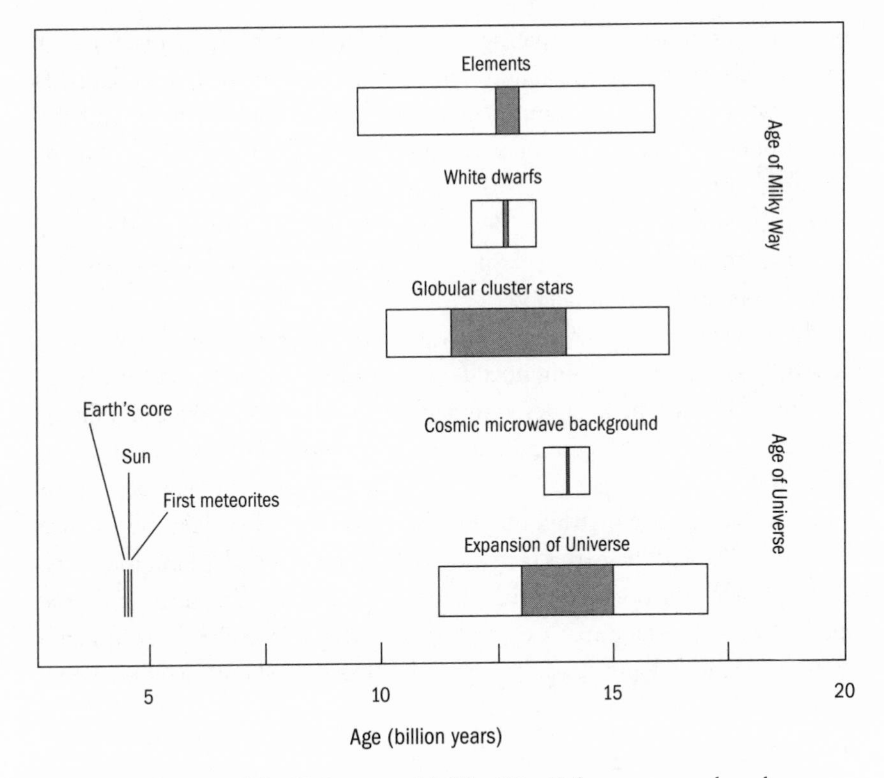

Figure 10.3 Ages of the Universe and Milky Way Galaxy compared to the ages of major dated events in the Solar System. Shaded areas include the range of calculated ages. Unfilled areas indicate the approximate uncertainties in the measurements.

ation by these two processes should be produced by the other known processes of nucleosynthesis.

Another argument for the original existence of the missing nuclides is that all of them are easily made in nuclear reactors. In general, they are no more difficult to make by artificial means than are many of the stable and long-lived radioactive nuclides. Considering the abundance and variety of natural nuclear reactors—the stars, supernovae, and other astrophysical events—it would be surprising indeed if the absent short-lived unstable nuclides were not present when the Solar System first formed.

Finally, there is concrete evidence that some of the missing nuclides did exist at one time. In Earth's atmosphere, for example, ^{129}Xe constitutes about one-fourth of total xenon. Yet in many meteorites, ^{129}Xe is as much as thirty times more abundant. Considering the very high probability that isotopes of the same element were thoroughly mixed when the Solar System formed, where did the excess ^{129}Xe come from? The most obvious answer, and the correct one, is that the ^{129}Xe was formed by the decay of ^{129}I (iodine), one of the short-lived missing nuclides that was incorporated into the meteorites when they formed but has since decayed out of existence. Some of the other extinct radioactive nuclides for which there is firm evidence of their past existence were discussed in Chapter 9.

This leaves the third and simplest hypothesis—that the elements were created long ago and the missing nuclides have simply decayed away over time. The abundance of the nuclides with half-lives slightly more than 82 million years and the half-lives of the missing nuclides are perfectly consistent with a Solar System whose age is 4.5 Ga, but inconsistent with an age much younger. Even an age of 1 billion years is insufficient to explain the absence of the nuclides with half-lives between 10 million years and 82 million years. Although it is not possible to calculate any precise age for Earth from the missing nuclides, their absence argues convincingly that the Solar System's age must be more than a few billion years but less than about 10 billion years.

A Consistent Age for the Universe

Despite the various uncertainties, it is quite satisfying that two very different methods of estimating the age of the Universe and three for estimating the age of the Milky Way Galaxy give results that are so consistent and mutually compatible (Figure 10.3). It seems clear that the Universe formed 13–15 billion years ago followed within 2 billion years or less by the formation of the Milky Way Galaxy. These ages are, in turn, consistent with the ages found for the meteorites, the Sun, and the formation of Earth's core, as well as the ages for the Moon's and Earth's oldest rocks. Thus, all the scientific data are, at present, self-consistent, suggesting that any errors that may exist in these various ages are relatively small.

It is all too easy to summarize the ages of the Universe, Milky Way Galaxy, and Solar System in a brief paragraph or a simple diagram. The evidence that led to this chronology, however, represents more than two centuries of clever research, and surely is one of the most notable and spectacular achievements of modern science.

CHAPTER ELEVEN

Summing Up and Looking Ahead

After more than three centuries of endeavor, scientists have concluded that the age of Earth is 4.5 billion years. This value is based on a great many age measurements using naturally occurring radioactive nuclides in meteorites and rocks from the Moon and Earth, and it has an uncertainty of less than 2%. In addition, the antiquity of Earth is consistent with independent evidence that the Milky Way Galaxy and the Universe are 11.5–14 and 13–15 billion years old, respectively.

Prior to the mid-eighteenth century, the age of Earth was based either in whole or in part on the interpretation of religious writings. The first attempt to find the age of Earth solely from scientific measurements and principles was by Benoit de Maillet in the early eighteenth century, who used the presumed decline of sea level to calculate an age of more than 2 billion years since the decline began.

During the nineteenth and early part of the twentieth centuries, four general types of methods were used to estimate the age of Earth. These included the cooling of Earth and the Sun, orbital physics, changes in ocean chemistry, and erosion and sedimentation. These methods relied on either the assumption of initial conditions now known to be incorrect, on the constancy of physical processes now known to be variable, or both. As a result, the methods produced a wide variety of incorrect values, ranging from a few million to billions of years.

Among the more important of these early age-of-Earth calculations were those of George Darwin, who in 1898 found a minimum age of 56 million years based on the increase in the orbital period of the Moon. Others include that of John Joly, who calculated an age of 89 million years based on the increase in the sodium content of the oceans (1899) and of Charles Walcott, who found 55 million years from the rate of accumulation of Paleozoic sedimentary rocks in western North America (1893). Probably the most influential calculations were those based on the time required for Earth to cool from a molten state and published by physicist Lord Kelvin and geologist Clarence King. Kelvin's first calculations, published in 1862, yielded a probable value of 98 million years. King refined Kelvin's method in 1893 and found 24 million years as the most probable age of Earth, a value that Kelvin subsequently endorsed.

There were dissenters. Scientists such as Charles Darwin and Thomas H. Huxley felt that the geologic and fossil record clearly indicated that much more time than 100 million years had passed since the first sedimentary rocks were deposited in ancient seas. Others, like John Perry and Thomas C. Chamberlin, challenged Kelvin's and King's basic assumptions and mathematics, and showed that their results were anything but exact. In the end, the dissenters proved to be correct. Methods based on cooling, orbital physics, the accumulations of chemicals in the oceans, and the thickness of sediments in the basins can never be made to yield a valid age for Earth because the initial conditions are uncertain and the rates of the processes involved have varied significantly over time.

The discovery of radioactivity by Antoine-Henri Becquerel in 1896 was the first in a series of developments that led to truly quantitative methods, based on the decay of long-lived radioactive nuclides that occur in rocks and minerals, from which a realistic age of Earth could be found. A few of the other milestones in this quest include:

—Suggestions by Ernest Rutherford in 1904 and 1905 that the accumulation of helium and lead might be used to find the age of uranium minerals.
—Demonstrations by Bertram Boltwood, Robert J. Strutt, and Arthur Holmes over the period 1905–1911 that uranium-bearing minerals contained radiogenic helium and lead and, therefore, that Rutherford's suggestion just might work.
—The idea advanced by Henry N. Russell in 1921 that a large portion of a planetary body, specifically Earth's crust, could be treated as a single geochemical reservoir and dated by the change in the composition of a radioactive parent-daughter pair.
—The development of the mass spectrometer beginning in 1914 with the parabola mass analyzer of James J. Thomson, who used it to dis-

cover isotopes, and culminating in the late 1930s with the development of the first modern mass spectrometer by Alfred O. Nier, who used his precision instrument to measure the variations of uranium and lead isotopes in lead ores.

—The idea advanced by E.K. Gerling in 1942 that the age of Earth could be calculated from the isotope composition of lead in lead ores, provided that the composition of primordial lead were known and assuming that the ores represent a fossilized sample of the lead composition of a single-stage reservoir within the Earth.

—The suggestion by Fritz G. Houtermans in 1947 that the isotope composition of primordial lead might be found in iron meteorites, and the implication that meteorites and the Earth were linked.

—The demonstration by Claire C. Patterson in 1956 that the isotope composition of lead in two iron meteorites, three stone meteorites, and modern Earth sediment all fall on a single Pb-Pb isochron indicating an age for the Earth-meteorite system of 4.55± 0.07 Ga.

—The development and refinement of modern instrumentation and radiometric dating methods beginning in the mid-1950s and continuing to the present.

These developments finally produced the answer to the question of the antiquity of Earth.

Radiometric Dating

Certain long-lived radioactive nuclides found in rocks and minerals serve as natural clocks that can be used to measure the ages of events in the history of Earth and the other solid bodies in the Solar System from which samples are available. The rates of decay of these radioactive parent nuclides have all been accurately measured in the laboratory and are known, from both experimental evidence and theoretical considerations, to be constant over the range of temperatures and pressures to which Earth, the Moon, and the meteorites have been subjected since formation. These radiometric dating systems can be partially or entirely reset, usually through loss of the daughter isotope, by thermal and chemical changes induced by metamorphism. This complicates but does not invalidate radiometric dating because:

—Post-crystallization metamorphism usually results in changes in rocks that can be recognized from their mineralogy, texture, chemical composition, or geologic setting, and this information is used to guide sample selection

—Most radiometric dating methods now in use are self-checking and provide information about the degree to which the isotope systems within the sample may or may not have been disturbed by post-crystallization events.

—Several of the radiometric methods, particularly U-Pb, Pb-Pb, and Ar-Ar, can be used to obtain valid crystallization ages from disturbed, as well as undisturbed, samples.

—Some methods, particularly Pb-Pb and Sm-Nd, are highly resistant to post-crystallization disturbances.

When properly applied, radiometric dating is an accurate and reliable method of determining the ages of rocks and sometimes of the complex events that have affected them since they formed.

The Age of Earth

The minimum age for Earth exceeds 4 billion years—the evidence is abundant and compelling. Rocks more than 3.5 billion years in age are found on all the continents. A number of these areas have been especially well studied, and the antiquity of their rocks is confirmed by hundreds of radiometric age measurements. In southern West Greenland, supracrustal rocks at Isua that were deposited in shallow Archean seas are 3.6–3.8 billion years old, as measured by U-Pb, Pb-Pb, Rb-Sr, and Sm-Nd methods. Younger granitoid gneisses that have Rb-Sr, Lu-Hf, U-Pb, and Pb-Pb ages of 3.5–3.7 billion years envelop the Isua supracrustal rocks. In the Pilbara block of western Australia volcanic rocks within a thick accumulation of supracrustal rocks have been dated as 3.5–3.6 billion years old by Rb-Sr, Sm-Nd, and U-Pb methods. These supracrustal rocks are intruded by granitoids with ages of 2.9–3.3 billion years. The Acasta Gneisses of the Slave Province in the Northwest Territories of Canada are currently the oldest known rocks on Earth. These rocks, originally granitoid plutons, have U-Pb ages as old as 4.03 billion years

There are other areas in the world where rocks exceeding 3.5 billion years in age have been found. These locations include the Superior Province of North America, Venezuela, the Liaoning Province of China, Zimbabwe, Swaziland, South Africa, Liberia, Sierra Leone, and Enderby Land in Antarctica.

The oldest terrestrial objects discovered so far, however, are not rocks but tiny zircon crystals that are found in younger sedimentary rocks in western Australia. Dated by U-Pb methods, they have ages ranging from 3.5 to 4.4 billion years. A part of one complex grain, in particular, crystallized 4.404± 0.008 billion years ago. The location of the source rocks from which these crystals were eroded is unknown, but the discovery of these ancient mineral grains offers hope that rocks whose ages are 4.4 billion years old or older may yet be found on Earth.

It is intriguing that most of the oldest rocks to date originated as sediments and lava flows, rather than as crystalline crust. Certainly, Earth's "primordial" crust, if such ever existed, has not yet been (and may never be) found, but the nature of the oldest rocks and minerals discovered thus far clearly indicates that Earth's age must exceed 4.4 billion years.

The Age of the Moon

In contrast to Earth, which has been a highly active planet since its formation, most of the significant events that have shaped the Moon occurred early in its history. As a result, nearly all Moon rocks are old by Earth standards, and the radiometric ages of nearly all of them exceed 3 billion years. Shortly after formation of the lunar crust, the Moon was subjected to bombardment by asteroid-sized objects that brecciated and fractured the crust to depths of as much as 20 km. These impacts also formed the large basins that dominate the topography of the lunar near side. Because of the early bombardment and the resultant disturbance of the isotope systems in the lunar rocks, only a handful of the returned lunar samples, all from the highlands, have radiometric ages of more than 4.2 billion years. Most of these are igneous cumulates, thought to have crystallized during the formation of the lunar crust. The oldest of these have Rb-Sr, U-Pb, and Sm-Nd ages in the range 4.42–4.56 Ga. Thus, the minimum age of the Moon must be 4.4–4.5 billion years.

The only viable hypothesis for the formation of the Moon is that it aggregated from the debris formed when a planetoid about the size of Mars or larger collided with a proto-Earth only about half its present size. If this hypothesis is correct, then the minimum age for the Moon is also a minimum age for Earth, and it follows that Earth and the Moon must be at least 4.5 billion years in age.

The Age of Meteorites

Meteorites are the oldest and most primitive rocks available for scientists to study. Their compositions and orbits indicate that nearly all meteorites are fragments of asteroids; the rare exceptions are from Mars and the Moon. Meteorites were freed from their parent bodies by collisions with other asteroids, and then thrown into Earth-crossing orbits by the gravitational field of Jupiter.

Meteorites vary widely in composition. Some, the chondrites, were formed by condensation and accretion of silicates from the Solar Nebula and are fragments of primitive asteroids. Others, like the achondrites, the stony irons, and the iron meteorites represent, respectively, surficial lava flows, the mantles, and the cores of asteroids that were large enough to undergo internal chemical differentiation.

Some meteorites were metamorphosed and their isotope systems disturbed by the collisions that fragmented their parent bodies. There are more than 100 well-dated meteorites, however, and the majority have individual Rb-Sr, Sm-Nd, Pb-Pb, and Ar-Ar ages of 4.4–4.6 billion years. Internal and whole-rock isochron ages measured by Rb-Sr, Lu-Hf, Sm-Nd, Pb-Pb, and Re-Os methods, as well as other isotope evidence, show that the major meteorite types were formed within a few millions years between 4.5 and 4.6 billion years ago.

All hypotheses for the formation of the Solar System propose that the planets, including Earth and the asteroids, formed within a very short interval of time after the condensation of silicates from the Solar Nebula. Thus, the ages of meteorites are relevant to the age of Earth, thereby suggesting that Earth and the other rocky bodies of the Solar System formed 4.5–4.6 billion years ago.

Lead Isotopes in Earth and the Meteorites

The isotope composition of lead when the matter of the Solar System first segregated into discrete bodies is preserved in a uranium-free mineral (troilite) found in iron meteorites. Internal Pb-Pb isochrons for a number of individual meteorites and for several meteorite groups all fall within the range of 4.4–4.6 billion years and pass through the primordial Pb composition found in the troilite of iron meteorites. The most precise meteorite ages are within the narrow range of 4.54–4.56 billion years.

The isotope compositions of lead in modern sediments and in young lead ores fall very close to the isochron formed by meteorites and the primordial lead composition. This suggests that meteorite and Earth leads evolved from a common, primordial lead isotope composition. The Pb-Pb growth curve formed by conformable lead ores and by several ancient leads passes through the lead isotope composition of troilite in iron meteorites but has an age of 4.49 billion years, which is probably the age of formation of Earth's core, rather than the age of the early proto-Earth.

Over the past decade, high-precision lead isotope measurements on selected meteorites, and measurements of isotopes produced by extinct radioactivities, have begun to reveal a detailed chronology for the early Solar System. These new data show that the parent bodies of the meteorites began to form 4.566 billion years ago, which is the Pb-Pb age of the calcium-aluminum inclusions in the Allende carbonaceous chondrite. Within only a few million years, the larger of these bodies had experienced melting and differentiation, and basalt lava flows erupted onto their surfaces. Analyses of meteorites from Mars indicate that it, too, formed early and only 10–15 million years later had developed a metallic core and ceased to grow.

It is likely that the proto-Earth began to grow at the same time as the meteorite parent bodies and Mars. The observation that terrestrial rocks contain the same tungsten isotope composition as the chondrite meteorites shows, however, that its core formed at least 50 million years later, probably after the formation of the Moon.

The Ages of the Galaxy and the Universe

The age of the Universe can be estimated in two ways. One utilizes the observed expansion of the Universe. The velocities at which galaxies are receding from the Milky Way and from each other can be measured, and the time required for the matter of the Universe to have expanded from a point of infinite density can then be calculated. The most recent calculations, which take into account the nonuniform expansion caused by the mass of the Universe and by the vacuum energy, a mysterious repulsive force that counteracts gravity, give ages for the Universe in the range of 13–15 billion years. Another method involves the analysis of the warmer and cooler spots in the cosmic microwave background, which gives an age for the Universe of 14.0 ± 0.5 billion years.

The age of the Milky Way Galaxy has been calculated in three ways. One method compares the observed stage of evolution of globular cluster stars with theoretical star evolution calculations. This method yields ages of 11.5–14.0 billion years. Another method is based on the present abundance and on the production rates in supernovae of ^{232}Th and ^{238}U, both of which are created by only one nucleosynthetic process. This method yields ages of 12.5–13.0 (± 3) billion years. Another method is based on the observed temperatures of white dwarfs in the globular cluster M4. Based on the time it had to take for these former stars to cool to their present temperatures, the Milky Way Galaxy must be at least 12.7 ± 0.7 billion years old.

The age of the Sun can be measured by the seismic waves, caused by turbulence in the Sun, that continuously propagate through its interior. These waves cause measurable disturbances on the Sun's surface and reveal its internal density structure, which is a function of a star's age. Recent analyses of these seismic waves give an age for the Sun of 4.5 ± 0.1 billion years. This value is in excellent agreement with the ages of meteorites, particularly since nuclear reactions began in the Sun after the formation of the meteorite parent bodies.

The ages found by various independent means for the Milky Way and the universe are older than, and therefore consistent with, the age of the Earth and Solar System found from the extensive radiometric dating of meteorites and rocks from the Moon and Earth. They are also consistent with the age of the Sun determined by analysis of the seismic waves that traverse its interior. Thus, the current data form a consistent picture that leaves no doubt that Earth and its cosmic surroundings are billions of years old.

Moving Forward

An important endeavor, and one that is being actively pursued, is the search for Earth's oldest rocks. No doubt this search will eventually result in the discovery of older rocks and minerals than have been found so far, but how much older, I cannot predict. The evidence for the existence of a lunar crust as early as 4.4–4.5 billion years ago suggests that Earth's crust also may have formed as early as 4.5 Ga. Earth, however, has been a tectonically active planet since birth, and it is also possible, if not probable, that Earth's earliest crust has long since been recycled.

Although the lead isotope growth curve for Earth, and the resulting age for Earth's core, is based on more than a dozen lead ores, it rests heavily on data from only four ancient leads. It is highly desirable that the data set of ancient leads be increased by a factor of two or three. Whether a significant number of additional lead ores with ages greater than 3 billion years will ever be found, however, is questionable.

We have much to learn about lunar history that is relevant to the age of Earth. In particular, too little is known about the nature and age of the early lunar crust, simply because there are too few samples and too little ground-based geologic information from the lunar highlands. It is quite likely that the age of the earliest lunar crust could be determined rather precisely given the right samples, but these samples are not in the existing lunar sample collections. The Moon is Earth's closest neighbor, however, and additional visits are neither impossible, impractical, nor prohibitively expensive. Remote landers with the capability of returning samples would help fill the gap somewhat, but better yet would be to send a few geologists. One of the severe shortcomings of the Apollo missions (Apollo 17 excepted) was the lack of geologic expertise among the crews. I hope NASA will not make the same mistake with future manned lunar missions.

Samples from the other planets and the asteroids would help clarify the early history of the Solar System immensely. Samples from Mars, Venus, and Mercury would all be of interest and, because of the expense and danger, could best be obtained by remote landers. Samples from Venus will be especially difficult to obtain because the excessive temperatures and pressures on its surface destroy spacecraft and their instruments rapidly.

The single most productive type of mission relevant to the age and earliest history of the Solar System, however, would return samples from one or more asteroids of chondrite composition, especially those similar to the primitive carbonaceous chondrites. Spectral studies indicate that asteroids with surface compositions resembling carbonaceous chondrites are common. Since carbonaceous chondrites and their parent bodies are thought to be among the most primitive bodies in the Solar System, samples undisturbed by massive collisions, by a stressful passage through Earth's atmosphere, and by contamination in Earth's environment would be very valuable. Analyses of such samples would further test the conclusion that asteroids are the sources of meteorites, and they would provide better and more accurate age and chemical information for the Solar System's most primitive objects. Perhaps the Japanese Institute of Space and Astronautical Science's MUSES-C mission to the Earth-crossing asteroid Itokawa (1998SF36), which was launched in May 2003 and will return a few grams of sample to Earth in 2007, will be the first in a series of missions designed to bring back samples of asteroids.

To supplement our knowledge about the age of Earth and the Solar System, we need refined estimates of the ages of the Milky Way Galaxy and the Universe. Spectacular progress has been made in the past decade, but more research is necessary. A more accurate age for the Milky Way will require refined models of star evolution and additional observations of the luminosity and temperature of globular cluster stars. It will also require more precise knowledge about the production history of the long-lived radioactive nuclides in supernovae and in other element-creating events in the Galaxy. Narrower limits still for the age of the Universe require much better ways to determine the distances to faraway galaxies, so that the Hubble constant and the resulting Hubble time can be measured more precisely. Determining the age of the Universe from its expansion also requires better values for the mass of the Universe and the nature and magnitude of the vacuum energy, so that the deceleration and acceleration history of the expansion can be estimated more accurately.

Despite the information that is lacking, the past five decades have produced a rich, consistent, and convincing body of evidence that the solid bodies of the Solar System formed 4.55 billion years ago and that the Universe has been in existence for 13–15 billion years. Our current understanding of the chronology and history of the Universe, our Galaxy, and the Solar System represents the fulfillment of a quest that has required more than two centuries of endeavor. Although our understanding will certainly be refined in the coming years, it already ranks as a wondrous triumph of modern science.

Recommended Reading

Albritton, C.C. Jr. 1980. *The Abyss of Time.* San Francisco: Freeman, Cooper and Company.

Burchfield, J.D. 1990. *Lord Kelvin and the Age of the Earth.* Chicago and London: University of Chicago Press.

Dalrymple, G.B. 1991. *The Age of the Earth.* Stanford, CA: Stanford University Press.

Hawking, S. 2001. *The Universe in a Nutshell.* New York: Bantam Books.

Lewis, C. 2000. *The Dating Game.* Cambridge and New York: Cambridge University Press.

Lewis, C.L.E., and S.J. Knell, eds. 2000. *The Age of the Earth: From 4004 B.C. to A.D. 2002.* London: Geological Society of London, Special Publication No. 190.

Lidsey, J.E. 2000. *The Bigger Bang.* Cambridge and New York: Cambridge University Press.

McSween, H.Y., Jr. 1987. *Meteorites and Their Parent Planets.* Cambridge and New York: Cambridge University Press.

Norton, O.R. 1994. *Rocks from Space.* Missoula, MT: Mountain Press.

Ridpath, I. 2001. *The Illustrated Encyclopedia of the Universe.* New York: Watson-Guptill.

Schopf, J.W. 1999. *Cradle of Life: The Discovery of Earth's Earliest Fossils.* Princeton, NJ: Princeton University Press.

Weinberg, S. 1993: *The First Three Minutes: A Modern View of the Origin of the Universe.* New York: Basic Books.

York, D. 1997. *In Search of Lost Time.* Bristol and Philadelphia: Institute of Physics Publishing.

References

Alibert, C., M.D. Norman, and M.T. McCulloch. 1994. An ancient Sm-Nd age for a ferroan noritic anorthosite clast from lunar breccia 67016. *Geochimica et Cosmochimica Acta,* vol. 58, pp. 2921–2926.

Allègre, C.J., G. Manhès, and C. Göepel. 1995. The age of the Earth. *Geochimica et Cosmochimica Acta,* vol. 59, pp. 1445–1456.

Appel, P.W.V., and S. Moorbath. 1999. Exploring Earth's oldest geologic record in Greenland. *EOS, Transactions of the American Geophysical Union,* vol. 80, pp. 257, 261, 264.

Bogard, D.D., and D.H. Garrison. 1995. ^{39}Ar-^{40}Ar age of the Ibitira eucrite and constraints on the time of pyroxene equilibration. *Geochimica et Cosmochimica Acta,* vol. 59, pp. 4317–4322.

Bogard, D.D., D.H. Garrison, and T.J. McCoy. 2000. Chronology and petrology of silicates from IIE iron meteorites: Evidence of a complex parent body evolution. *Geochimica et Cosmochimica Acta,* vol. 64, pp. 2133–2154.

Boltwood, B.B. 1907. On the ultimate disintegration products of the radioactive elements. Part II. The disintegration products of uranium. *American Journal of Science,* series 4, vol. 23, pp. 77–88.

Borg, L., M. Norman, L. Nyquist, G. Snyder, L. Taylor, M. Lindstrom, and H. Weismann. 1997. A relatively young samarium-neodymium age of 4.36 Ga for ferroan anorthosite 62236. *Meteoritics and Planetary Science,* vol. 32, p. A18.

Brouxel, M., and M. Tatsumoto. 1991. The Estherville mesosiderite: U-Pb, Rb-Sr, and Sm-Nd isotopic study of a polymict breccia. *Geochimica et Cosmochimica Acta,* vol. 55, pp. 1121–1133.

Carlson, R.W., and G.W. Lugmair. 1988. The age of ferroan anorthosite 60025: Oldest crust on a young moon. *Earth and Planetary Science Letters,* vol. 90, pp. 119–130.

Chamberlin, T.C. 1899. On Lord Kelvin's address on the age of the Earth as an abode fitted for life. *Annual Report of the Smithsonian Institution, 1899*, pp. 223–246.

Chen, J.H., D.A. Papanastassiou, and G.J. Wasserburg. 1998. Re-Os systematics in chondrites and the fractionation of the platinum group elements in the early solar system. *Geochimica et Cosmochimica Acta*, vol. 62, pp. 3379–3392.

Doe, B.R., and J.S. Stacey. 1974. The application of lead isotopes to the problems of ore genesis and ore prospect evaluation: A review. *Economic Geology*, vol. 69, pp. 757–776.

Freedman, W.H. 2000. The Hubble constant and the expansion age of the Universe. *Physics Reports*, vol. 333-334, pp. 13–31.

Gilkey, L. 1985. *Creationism on Trial.* Minneapolis, MN: Winston Press.

Göpel, C., G. Manhès, and C.J. Allègre. 1994. U-Pb systematics of phosphates from equilibrated ordinary chondrites. *Earth and Planetary Science Letters*, vol. 121, pp. 153–171.

Halliday, A.N., and D.-C. Lee. 1999. Tungsten isotopes and the early development of the Earth and Moon. *Geochimica et Cosmochimica Acta*, vol. 63, pp. 4157–4179.

Hartmann, W.K. 1983. *Moons and Planets*, 2nd ed. Belmont, CA: Wadsworth.

Holmes, A. 1911. The association of lead with uranium in rock-minerals, and its application to the measurement of geological time. *Proceedings of the Royal Society of London*, series A, vol. 85, pp. 248–256.

Holmes, A. 1927. *The Age of the Earth: An Introduction to Geological Ideas.* London: Ernest Benn (Benn's Sixpenny Library, No. 102).

Horan, M.V., J.W. Morgan, R.J. Walker, and J.H. Grossman. 1992. Rhenium-osmium isotope constraints on the age of iron meteorites. *Science*, vol. 255, pp. 1118–1121.

Johnson, H.L., and A.R. Sandage. 1956. Three-color photometry in the globular cluster M3. *Astrophysical Journal Supplement*, vol. 124, pp. 379–389.

Lugmair, G.W., and S.J.G. Galer. 1992. Age and isotopic relationships among the angrites Lewis Cliff 86010 and Angra dos Reis. *Geochimica et Cosmochimica Acta*, vol. 56, pp. 1673–1694.

Manhès, G., J.-F. Minster, and C.J. Allègre. 1978. Comparative uranium-thorium-lead and rubidium-strontium study of the Saint Severin amphoterite: Consequences for early Solar System chronology. *Earth and Planetary Science Letters*, vol. 39, pp. 14–24.

McCoy, T.J., K. Keil, R.N. Clayton, T.K. Mayeda, D.D. Bogard, D.H. Garrison, G.R. Huss, I.D. Hutcheon, and R. Wieler. 1996. A petrologic, chemical and isotopic study of Monument Draw and comparison with other acapulcoites: Evidence for formation by incipient partial melting. *Geochimica et Cosmochimica Acta*, vol. 60, pp. 2681–2708.

McCulloch, M.T. 1996. Isotopic constraints on the age and early differentiation of the Earth. *Journal of the Royal Society of Western Australia*, vol. 79, pp. 131–139.

Minster, J.F., and C.J. Allègre. 1979. ^{87}Rb-^{87}Sr chronology of H chondrites: Constraint and speculations on the early evolution of their parent body. *Earth and Planetary Science Letters*, vol. 42, pp. 333–347.

Minster, J.F., J.L. Birck, and C.J. Allègre. 1982. Absolute age of formation of chondrites studied by the ^{87}Rb-^{87}Sr method. *Nature*, vol. 300, pp. 414–419.

Moorbath, S., M. J. Whitehouse, and B. S. Kamber. 1997. Extreme Nd-isotope heterogeneity in the early Archean—fact or fiction? Case histories from northern Canada and West Greenland. *Chemical Geology*, vol. 135, pp. 213–231.

Morgan, J.W., R.J. Walker, and J.N. Grossman. 1992. Rhenium-osmium isotope systematics in meteorites: I, Magmatic iron meteorite groups IIAB and IIIAB. *Earth and Planetary Science Letters*, vol. 108, pp. 191–202.

Niemeyer, S. 1979. ^{40}Ar-^{39}Ar dating of inclusions from IAB iron meteorites. *Geochimica et Cosmochimica Acta*, vol. 43, pp. 1829–1840.

Norman, M., L. Borg, L.E. Nyquist, D.D. Bogard, G. Snyder, L. Taylor, and M. Lindstrom. 1998. Composition and age of the lunar highlands: Petrogenesis of ferroan noritic anorthosite 62236. Lunar and Planetary Science Conference 29th.

Nutman, A.P., V.R. McGregor, C.R.L. Friend, V.C. Bennett, and P.D. Kinney. 1996. The Itsaq Gneiss Complex of southern West Greenland: The world's most extensive record of early crustal evolution (3900–3600 Ma). *Precambrian Research*, vol. 78, pp. 1–39.

Nyquist, L.E., D.D. Bogard, H. Wiesmann, B.M. Bansal, C-Y. Shih, and R.M. Morris. 1990. Age of a eucrite clast from the Bholghati howardite. *Geochimica et Cosmochimica Acta*, vol. 54, pp. 2195–2206.

Patterson, C.C. 1956. Age of meteorites and the earth. *Geochimica et Cosmochimica Acta*, vol. 10, pp. 230–237.

Pellas, P., C. Fieni, M. Trieloff, and E.K. Jessberger. 1997. The cooling history of the Acapulco meteorite as recorded by the ^{244}Pu and ^{40}Ar-^{39}Ar chronometers. *Geochimica et Cosmochimica Acta*, vol. 61, pp. 3477–3501.

Premo, W.R., and M. Tatsumoto. 1991. U-Th-Pb isotopic systematics of lunar norite 78235. *Proceedings of the 21st Lunar and Planetary Science Conference*, pp. 89–100.

Premo, W.R., and M. Tatsumoto. 1992. U-Th-Pb, Rb-Sr, and Sm-Nd isotopic systematics of lunar troctolitic cumulate 76535: Implications on the age and origin of this early lunar, deep-seated cumulate. *Proceedings of the 22nd Lunar and Planetary Science Conference*, pp. 381–397.

Premo, W.R., M. Tatsumoto, K. Misawa, N. Nakamura, and N.I. Kita. 1999. Pb-isotope systematics of lunar highland rocks (>3.9 Ga): Constraints on early lunar evolution. In *Planetary Petrology and Geochemistry (Lawrence A. Taylor Volume)*, pp. 207–240. Columbia, MO: Bellwether Publishing.

Prinzhofer, A., D.A. Papanastassiou, and G.J. Wasserburg. 1992. Samarium-neodymium evolution of meteorites. *Geochimica et Cosmochimica Acta*, vol. 56, pp. 797–815.

Renne, P.R. 2000. $^{40}Ar/^{39}Ar$ age of plagioclase from Acapulco meteorite and the problem of systematic errors in cosmochronology. *Earth and Planetary Science Letters*, vol. 175, pp. 13–26.

Russell, H.N. 1921. A superior limit to the age of the Earth's crust. *Proceedings of the Royal Society of London*, series A, vol. 99, pp. 84–86.

Rutherford, E. 1905. Present problems in radioactivity. *Popular Science Monthly*, vol. 67, pp. 5–34.

Rutherford, E., and F. Soddy. 1903. Radioactive change. *Philosophical Magazine*, series 6, vol. 5, pp. 576–591.

Shen, J.J., D.A. Papanastassiou, and G.J. Wasserburg. 1996. Precise Re-Os determinations and systematics of iron meteorites. *Geochimica et Cosmochimica Acta*, vol. 60, pp. 2887–2900.

Shen, J.J., D.A. Papanastassiou, and G.J. Wasserburg. 1998. Re-Os systematics in pallasite and mesosiderite metal. *Geochimica et Cosmochimica Acta*, vol. 62, pp. 2715–2723.

Shih, C.Y, L.E. Nyquist, E.J. Dasch, D.D. Bogard, B.M. Bansal, and H. Weismann. 1993. Ages of pristine norite clasts from lunar breccias 15445 and 15455. *Geochimica et Cosmochimica Acta*, vol. 57, pp. 915–931.

Shu, F.S. 1982. *The Physical Universe: An Introduction to Astronomy*. Mill Valley, CA: University Science Books.

Smolier, M.I., R.J. Walker, and J.W. Morgan. 1996. Re-Os ages of group IIA, IIIA, IVA, and IVB iron meteorites. *Science*, vol. 271, pp. 1099–1102.

Stewart, B., D.A. Papanastassiou, and G.J. Wasserburg. 1996. Sm-Nd systematics of a silicate inclusion in the Caddo IAB iron meteorite. *Earth and Planetary Science Letters*, vol. 143, pp. 1–12.

Takeda, H., D.D. Bogard, D.W. Mittlefehldt, and D.H. Garrison. 2000.

Mineralogy, petrology, chemistry, and ^{39}Ar-^{40}Ar and exposure ages of the Caddo County IAB iron: Evidence for early partial melt segregation of a gabbro area rich in plagioclase-diopside. *Geochimica et Cosmochimica Acta*, vol. 64, pp. 1311–1327.

Tatsumoto, M., D.M. Unruh, and G.A. Desborough. 1976. U-Th-Pb and Rb-Sr systematics of Allende and U-Th-Pb systematics of Orgueil. *Geochimica et Cosmochimica Acta*, vol. 40, pp. 617–634.

Tera, F., R.W. Carlson, and N.Z. Boctor. 1997. Radiometric ages of basaltic achondrites and their relation to the early history of the solar system. *Geochimica et Cosmochimica Acta*, vol. 61, pp. 1713–1731.

Thomson, W. (Lord Kelvin) 1864. On the secular cooling of the Earth. *Royal Society of Edinburgh Transactions*, vol. 23, pp. 157–159.

Turner, G., P.H. Cadogan, and C.J. Younge. 1978. The early history of chondrite parent bodies inferred from ^{40}Ar-^{39}Ar ages. *Proceedings of the Ninth Lunar and Planetary Science Conference*, pp. 989–1025.

Upham, W. 1893. Estimates of geologic time. *American Journal of Science*, 3rd series, vol. 45, pp. 209–220.

Wilde, S.A., and J.W. Valley. 2001.Evidence from detrital zircons for the existence of continental crust and oceans on the Earth 4.4 Gyr ago. *Nature*, vol. 409, pp. 175–178.

York, D., and G.B. Dalrymple. 2000. Comments on a creationist's irrelevant discussion of isochrons. *Reports of the National Center for Science Education*, vol. 20, pp. 18–20, 25–27.

Glossary

abscissa

The horizontal axis of a graph.

achondrite

A type of meteorite that lacks chondrules.

Age of Enlightenment

A period of European history during the seventeenth and eighteenth centuries when scientific and intellectual developments fostered a belief in natural law and confidence in conclusions reached by observation and reason.

alpha decay

Radioactive decay in which an alpha particle (helium nucleus) is ejected from the nucleus of the atom.

alpha particle

A helium nucleus consisting of two protons and two neutrons.

amino acid

A family of twenty organic molecules that form proteins. The sequence of the amino acids in a protein determine the protein's function.

anaerobic

The ability to live in the absence of free oxygen.

andesite

A fine-grained volcanic rock that contains plagioclase feldspar and one or more of the mafic minerals pyroxene, hornblende, and biotite. Andesite is higher in silica and less mafic than basalt. It is a common volcanic rock erupted from the volcanoes along active continental margins, such as the Cascade Mountains of North America.

angular momentum

A measure of the intensity of rotational motion.

anorthosite

A rock composed almost entirely of plagioclase feldspar.

asteroid

A small rocky planet. Asteroid diameters range from about 1 meter to 1000 km. Most asteroids orbit the Sun in a belt between Mars and Jupiter.

asteroid belt

The locus of asteroids that orbit the Sun between Mars and Jupiter.

atmosphere

The pressure equivalent to that of Earth's atmosphere at sea level, or about 14.7 pounds per square inch.

banded iron formation

A chemical sedimentary rock consisting of layers of iron-rich and iron-poor chert; a source of iron ore.

basalt

A fine-grained volcanic rock composed primarily of plagioclase feldspar, olivine, and pyroxene. Basalt is a common rock on Earth and the other solid bodies of the Solar System. The floors of the oceans, Hawaiian volcanoes, the lunar maria, and the surfaces of some asteroids are composed primarily of basalt.

basaltic achondrite

An achondrite meteorite with a composition similar to basalt.

basin

A large, multi-ringed, impact crater. Lunar basins are 200–2000 km in diameter.

batholith

A large body of igneous rock that intrudes the crust and cools beneath the surface.

beta decay

A type of radioactive decay in which a beta particle (an electron) is ejected from the nucleus of an atom, resulting in a neutron being converted into a proton.

beta particle

An electron that originates in the nucleus of an atom.

Big Bang

The scientific theory that the universe originated in a primeval "explosion" from an initial state of extremely high density and has been expanding, or inflating, ever since.

biotite

A dark-colored mica mineral.

block

See craton.

breccia

A coarse-grained fragmental rock composed primarily of angular, broken rock fragments. Breccia may originate by either volcanic or sedimentary processes.

calcium-aluminum inclusion (CAI)

A tiny inclusion of unusual high-temperature minerals found in some meteorites. CAIs are thought to be the first solid objects formed in the Solar Nebula.

carbonaceous chondrite

A chondrite meteorite with up to 5% of carbon, including organic molecules.

Celsius (°C)

A temperature scale in which the boiling point of water at 1 atmosphere of pressure is 100° and the freezing point is 0°; named after the Swedish astronomer Anders Celsius, who established the scale in 1742.

Cepheid

A type of pulsating star whose variations in luminosity are periodic.

chemical sedimentary rock

A rock formed by the precipitation of material from water.

chert

An extremely fine-grained sedimentary rock consisting primarily of silica. Chert is formed either from the remains of siliceous microorganisms or by direct precipitation of silica from sea water.

chloride

A chemical compound that contains chlorine. Table salt ($NaCl$) is a chloride.

chondrite

A family of meteorites that contain chondrules.

chondrule

A small spherical inclusion of silicate minerals, principally olivine and pyroxene, found in chondrite meteorites.

closed system

A system in which matter neither enters nor leaves. A system may be very small, like a mineral grain, or very large, like a galaxy.

comet

A small body, consisting mostly of ices, that orbits the Sun. Most comets are thought to come from beyond the orbits of Pluto and Neptune.

concordia

The locus of concordant U-Pb ages on a $^{207}Pb/^{235}U$ vs. $^{206}Pb/^{238}U$ diagram.

conduction
: The transmission of energy, such as heat, through a substance without movement of the substance itself. *See also* convection.

conformable lead
: A lead ore deposit that conforms to the geometry of the sedimentary beds that enclose it. Conformable ores are thought to be chemical sedimentary deposits.

conglomerate
: A coarse-grained sedimentary rock consisting of rounded rocks or pebbles cemented together by finer-grained material.

conservation of momentum
: A scientific law governing interactions between two or more bodies. It requires that the total momentum of the bodies remains constant in time, absent outside influences.

convection
: Mass movement in a fluid in a gravitational field because of differences in temperature. Convection moves both the fluid and the heat (energy). *See also* conduction.

Cordilleran Sea
: A shallow inland sea that occupied much of western North America during the Paleozoic era.

core
: The central part of a planet or asteriod. Earth's core begins about 2900 km below the surface and is probably composed of nickel and iron, with an outer liquid part and an inner solid part.

cosmic dust
: Tiny particles of solid matter that occupy interstellar space.

cosmic microwave background
: Radiation that originated in the Big Bang and has been cooling ever since. It is now at a temperature of 2.7 Kelvin, as predicted by the Big Bang theory.

cosmic-ray exposure age
: The time that a rock has been exposed to cosmic rays, calculated from isotope changes induced by the cosmic rays over the time of the exposure.

cosmological constant
: A term inserted by Einstein into the equations for relativity that accounts for the vacuum energy (a repulsive force) of the Universe.

cosmologist
: A scientist who works in the field of cosmology.

cosmology
: The study of the origin, structure, and history of the Universe.

craton
: The relatively stable core of a continent, usually of Precambrian age. Also called shield or block.

crust
: The outermost part of Earth or any planet. Earth's continental crust is about 30–50 km thick, whereas the oceanic crust is about 5–10 km thick.

cumulate
: An igneous rock consisting of crystals that have been concentrated by either sinking or floating in the parent magma.

cyanobacteria
: A group of microorganisms capable of producing oxygen by photosynthesis. Also called blue-green algae.

dacite
: A volcanic rock consisting primarily of plagioclase, quartz, hornblende, or pyroxene. Dacite is higher in silica and more felsic than andesite.

daughter nuclide
: A nuclide produced by radioactive decay of a parent nuclide.

decay constant
: The probability that a radioactive nuclide will decay in a given period of time.

deformation
: A change in the form or volume of a rock mass due to tectonic forces. Rocks are deformed most commonly by faulting, folding, and plastic flow.

detrital sedimentary rock
: A sedimentary rock composed of grains eroded from pre-existing rocks.

diabase
: A shallow intrusive rock with a basaltic composition and a distinctive texture. Diabase commonly occurs as shallow, tabular intrusions.

differentiation
: The process by which igneous rocks of different compositions are formed from a single parent magma. The most common mechanisms of differentiation are fractional crystallization and the concentration of crystals by sinking or floating.

dike

A tabular body of igneous rock that cuts across the bedding or structure of the rocks it intrudes.

discordia

A straight line on a $^{207}Pb/^{235}U$ vs. $^{206}Pb/^{238}U$ diagram that connects two points on concordia, the original age of the rock with the time it was disturbed.

Doppler effect

The change in wavelength and frequency as a source of electromagnetic radiation or sound moves toward or away from an observer.

dunite

A cumulate igneous rock consisting primarily of the mineral olivine.

ecliptic plane

The imaginary plane that contains Earth's orbit and extends infinitely outward in all directions.

effective temperature

The surface temperature a star would have, based on its color, if it were a perfect radiator of energy (a blackbody).

electromagnetic radiation

Any form of radiation, including visible and invisible light, radio waves, X-rays, etc.

electron

A negatively charged particle of low mass that resides in the outer electron shells of an atom.

electron capture

A type of radioactive decay in which an electron from the innermost electron shell falls into the nucleus and a proton is converted to a neutron. The result is a different element of the same mass.

Enlightenment

See Age of Enlightenment.

enstatite

A type of pyroxene mineral composed of magnesium, silicon, and oxygen.

enstatite chondrite

A type of chondrite meteorite composed mostly of the mineral enstatite.

evolution

The scientific theory that all types of living organisms have their origin in pre-existing organisms and change by modifications through successive generations. Evolution is the grand unifying theory of biology. Also known as descent with modification.

exponential equation

An equation that contains the term e^x, where e is the base of natural logarithms (e = 2.71828+) and x is any other number. Radioactive decay and population growth are two phenomena that can be described by exponential equations and are, therefore, said to be exponential.

Fahrenheit (°F)

A temperature scale in which the boiling point of water at 1 atmosphere of pressure is 212° and the freezing point is 32°; named after the German-Dutch physicist G.D. Fahrenheit, who established the scale in 1724.

feldspar

A mineral group common in Earth's crustal rocks. Feldspars are light-colored, silicate minerals that contain aluminum and oxygen, and also may contain sodium, potassium, and calcium as major elements. Plagioclase is a type of feldspar.

felsic rock

A qualitative term for light-colored rocks. Felsic rocks contain an abundance of light-colored minerals, such as feldspar and quartz, and a low proportion of dark minerals, such as pyroxene, hornblende, and olivine. *See also* mafic rock.

fergusonite

A yttrium oxide mineral. It may also contain niobium, tantalum, iron, titanium, and uranium.

flat universe

A universe in which Euclidean geometry applies.

Ga

Giga annum, or billion (10^9) years.

gabbro

A plutonic rock consisting primarily of plagioclase feldspar and pyroxene; the intrusive equivalent of basalt.

galactic disk

The flattened, disk-shaped part of a spiral galaxy.

galactic plane

The plane of rotation of a disk-shaped galaxy.

galaxy

An aggregation of stars, numbering in the billions, that rotate about a common center. Many galaxies are disk-shaped.

galena

A mineral composed of lead sulfide.

geochron

A special isochron on a Pb-Pb isotope diagram that indicates the age of meteorites (the meteoritic geochron) or Earth (the terrestrial geochron).

geothermal gradient

The change in temperature as a function of depth within Earth.

globular cluster

A roughly spherical group of old stars, numbering 100,000 or more, that resides in a halo surrounding the center of the Galaxy. The stars in globular clusters are thought to be among the oldest stars in the Galaxy.

gneiss

A coarse-grained metamorphic rock in which layers rich in granular minerals alternate with layers rich in platy minerals. Pronounced "nice."

graphite

A form of pure carbon.

granite

An igneous rock consisting mostly of potassium feldspar and quartz.

granitoid

A general term for felsic and intermediate plutonic rocks. Granite, diorite, and tonolite, to name a few specific rock types, are granitoids.

greenhouse effect

The trapping of heat by clouds or certain chemicals (such as carbon dioxide) that allows heat to reach the surface of a planet but does not allow all the heat trying to escape to do so. The result is a warming of the planet. Venus is hot because of the greenhouse effect, owing to the abundance of carbon dioxide in its atmosphere.

half-life

The time it takes for one-half of any quantity of a radioactive nuclide to decay.

Hertzsprung-Russell (H-R) diagram

A graph of the luminosity versus the effective temperature of stars.

highlands

The heavily cratered upland areas of the moon.

hornblende

A dark-colored silicate mineral with calcium, sodium, magnesium, iron, aluminum, and oxygen as principal constituents.

Hubble constant

A number (H_0) that specifies the change in velocity with distance in the expanding Universe.

Hubble flow

The outward movement or expansion of all galaxies in the Universe away from each other.

Hubble's law

The relationship between velocity and distance in the expanding Universe.

Hubble time

The age of the Universe determined by inversion of the Hubble constant (equals $1/H_0$).

hypothesis

A tentative model, to be tested, that may explain a set of incomplete observations.

igneous rock

A rock formed by the crystallization of molten rock.

ilmenite

An iron titanium oxide mineral that is a minor but common constituent of igneous rocks.

impact crater

A crater formed by the impact of a meteorite.

inclusion

An identifiable body of rocks; a rock, or a mineral grain, that is enclosed within another rock. Usually inclusions are older than the enclosing rock.

inductive reasoning

The logical process of drawing general conclusions from specific observations.

inert gas

A gaseous element whose outer electron shells are filled, so it generally does not combine chemically with other elements. Inert gases include helium, neon, argon, krypton, and xenon. Also called noble or rare gases.

infrared light

Light with wavelengths that lie outside of the red end of the visible light spectrum.

intrusive rock

A rock that crystallized from a magma after the magma was injected into another body of rock.

ion

An atom with either an excess or a deficiency of electrons, which results in a positive or negative electrical charge.

ion probe

A type of mass spectrometer that uses a beam of ions to blast atoms off the surface of a substance so that their isotope compositions can be measured.

iron meteorite

A meteorite composed primarily of nickel-iron alloy.

isochron

A line of equal age on an isotope correlation diagram. The slope of an isochron is a function of age.

isochron age

The age determined from an isochron.

isotope

A nuclide of an element that differs from other nuclides of the same element in the number of neutrons contained in the nucleus. The chemical behavior of all isotopes of an element is identical.

isotope dating

See radiometric dating.

Kelvin (K)

A temperature scale named after Lord Kelvin. One Kelvin equals 1° Celsius, but the Kelvin scale has its zero-point at absolute zero, which is –273.16° C. Absolute zero is the temperature at which all molecular motion ceases; a body at that temperature has no heat energy. By convention, the word "degrees" and its symbol are not used when expressing temperature in Kelvin.

lava

Molten rock that erupts onto the surface of a planet or asteroid.

law, scientific

A concise statement of a simple relationship between phenomena that is invariable under a given set of conditions.

light year

The distance light travels in vacuum in 1 year, or 9.46×10^{12} km.

limestone

A sedimentary rock consisting primarily of calcium carbonate. Most limestones are formed by the accumulation of biogenic remains (shells) or by chemical precipitation from sea water.

luminosity

A measure of the brightness of a star, usually expressed relative to the Sun.

Ma

Mega annum, or million (10^6) years.

mafic rock

A qualitative term for dark-colored rocks. Mafic rocks contain an abundance of the dark-colored minerals, such as hornblende, pyroxene, and olivine, and a relatively lower proportion of light-colored minerals, such as quartz and feldspar. *See also* felsic rock.

main sequence

The band on a Hertzsprung-Russell diagram in which all stars begin their lives.

magma

Molten rock.

mantle

The portion of Earth between the crust and core.

mare

A large, smooth lava plain, usually occupying the floor of an impact basin on the Moon.

maria

Plural form of mare.

mass number

The total number of protons and neutrons in a nuclide.

mass spectrograph

A forerunner of the mass spectrometer, which used a photographic plate as a detector.

mass spectrometer

An instrument for measuring the relative amounts of various nuclides according to their mass.

megaparsec

One million parsecs.

melt rock

A rock that has recrystallized from a rock melt created by an energetic impact.

metamorphic rock

A rock formed by metamorphism.

metamorphism

The process by which the composition, texture, or structure of rocks is altered in the solid state due to heat, pressure, and the introduction of new chemical constituents. Igneous, sedimentary, or metamorphic rocks may undergo metamorphism and thereby become metamorphic rocks.

meteor

The streak of light caused when a meteoroid enters Earth's atmosphere.

meteorite

A meteoroid that has fallen to Earth.

meteoritic dust

Very fine-grained meteoroid material.

meteoroid

A rock fragment from an asteroid or another planet before it reaches Earth.

mica

A group of platy minerals, the most common of which is biotite.

micrometeorite

Small particles of meteoroid material, some of which may come from comets.

Milky Way Galaxy

The galaxy in which the Solar System resides; sometimes simply called the Galaxy.

model age

An isotope age based on two data points, as opposed to an isochron age, in which three or more data points form an isochron.

momentum

A measure of the intensity of motion and the resistance of an object to changes in its direction or velocity.

multi-stage lead

A lead that has evolved isotopically in two or more systems with different U/Pb ratios. *See also* single-stage lead.

nebula

A body of interstellar gas and dust.

neutron

An electrically neutral particle found in the nucleus of all nuclides except ^{1}H (hydrogen one) and with a mass very close to that of a proton.

norite

A variety of gabbro.

nucleosynthesis

A general term for the processes by which elements are created.

nucleus

The core of an atom, containing protons and neutrons. Most of the mass and available energy of an atom reside in the nucleus.

nuclide
An atom with a distinct number of protons and neutrons. Nuclides include all isotopes of all elements.

olivine
An iron-magnesium silicate that is common in mafic and ultramafic igneous and metamorphic rocks.

ordinary chondrite
A chondrite meteorite that does not contain chondrules.

ordinate
The vertical axis of a graph.

parent body
The source of a meteorite, usually an asteroid but also the Moon and Mars.

parent nuclide
Any radioactive nuclide.

parsec
The distance at which the mean radius of Earth's orbit subtends an angle of 1 second of arc. It is equal to 3.26 light years, or 3.08×10^{13} km.

partial melting
The process of forming magma from a parent rock by selective melting of some of the constituents. The selective melting is controlled by temperature, pressure, and composition.

phosphate mineral
A mineral containing phosphorous, in the form PO_4, as a primary element.

photosynthetic organism
An organism that creates oxygen by the process of photosynthesis. Plants and cyanobacteria are examples.

pillow lava
A lava that exhibits pillow structure. The pillows are rounded masses of chilled rock that form only in lava that flows and cools under water.

plagioclase
A calcium-sodium feldspar and one of the most common rock-forming minerals.

plane of the ecliptic
See ecliptic plane.

planetesimal
A small rocky body orbiting the Sun from which the planets aggregated.

plate tectonics

The scientific theory that describes the configuration, movement, and structure of the lithospheric plates that constitute Earth's outermost shell, as well as the phenomena caused by the plate motions.

pluton

A large body of igneous rock that crystallized beneath the surface.

plutonic rock

A rock formed in a pluton.

ppm

Parts per million.

primordial lead

Lead with the same isotope composition as when the Solar System formed.

protein

A group of organic molecules, each built from a long chain of amino acids, that determine the unique shapes and functions of the cells in living organisms.

proto-Earth

Earth before it reached its present size.

proto-galaxy

A galaxy in the process of forming.

proto-planet

A planet in the process of forming.

proton

An electrically positive particle found in the nucleus of all elements and with a mass very close to that of a neutron.

pyroxene

A mineral group whose principal constituents, in addition to oxygen, are usually iron, magnesium, calcium, aluminum, and silicon. Pyroxene is a common mineral in igneous and metamorphic mafic and ultramafic rocks.

quartz

A mineral formed of silicon dioxide.

radiogenic

A nuclide created by radioactive decay.

radiometric dating

The process of measuring the age of a rock using naturally occurring radioactive isotopes. Also called isotope dating.

rare earth element

The group of elements with atomic numbers from 57 (lanthanum) to 71 (lutetium).

red giant

A very large star with a relatively low temperature that has exhausted its hydrogen fuel. The Sun will become a red giant in about 5 billion years.

red shift

The change in the color (wavelength) of light toward redder colors as the light source moves away from the observer; due to the Doppler effect.

reflectance spectrophotometry

A technique for measuring the wavelengths of light reflected from the surface of a body; used to indicate surface composition.

regolith

The lunar "soil," composed primarily of pulverized rock and glass.

relativity

See theory of relativity.

rhyolite

The volcanic equivalent of granite.

r-process

The process by which new nuclides are created from existing nuclides through a very high flux of neutrons. The r-process (r for rapid) occurs in astronomical catastrophic events, such as supernovae. *See also* s-process.

sand

Detrital material in the size range of 0.0625–2.0 mm.

sandstone

A rock composed of sand grains.

schist

A metamorphic rock with a laminate structure caused by the subparallel alignment of mica, which is a principal constituent.

sedimentary rock

A rock formed by either the deposition of detrital grains, including fragments of the shells of animals, or the chemical precipitation of material directly from sea water.

shale

A sedimentary rock consisting primarily of clay-sized (microscopic) particles.

shield

See craton.

siderophile

A term applied to elements whose chemical behavior is similar to that of iron.

silicate

A mineral that contains silicon dioxide as a principal constituent.

single-stage lead

A lead that has evolved isotopically in only one system with one initial U/Pb ratio since the Solar System formed. *See also* multi-stage lead.

Solar Nebula

The rotating disk-shaped cloud of dust and gas from which the Solar System formed.

Solar System

The star system that includes the Sun and Earth.

solar wind

The outflow of charged particles from the Sun.

s-process

The process by which new nuclides are created through a relatively low flux of neutrons. The s-process (s for slow) is thought to occur primarily in red giant stars. *See also* r-process.

stone meteorite

A meteorite composed primarily of silicate minerals.

stony iron meteorite

A meteorite composed of silicate minerals and iron-nickel alloy.

stratigraphy

The study of the composition, formation, sequence, and correlation of sedimentary (stratified) rocks.

sulfate

Any compound containing the sulfur tetraoxide molecule (SO_4).

sulfide

Any compound containing sulfur and one other element as a cation. The mineral troilite (FeS) is a sulfide.

supernova

The explosion of a star.

supracrustal rock

A sedimentary rock or lava flow that was deposited on Earth's crust.

terrane

The area over which a particular group of rocks is prevalent.

terrestrial
Pertaining to Earth or land.

theory of evolution
See evolution.

theory of relativity
The scientific theory that describes physical phenomena observed from frames of reference that are moving with respect to each other. The special theory deals with constant relative motion in a straight line. The general theory deals with accelerating relative motion. Relativity is the unifying theory for the physics of space and time.

theory, scientific
A logical and unifying structure of ideas that accounts for a large body of observations and thus explains something important about nature.

thermal conductivity
A measure of the ability of a material to conduct heat.

thermodynamics
The physics and mathematical treatment of heat and its relation to chemical and physical phenomena.

tidal friction
Friction in a planetary body caused by the gravitational pull of an orbiting body. The Moon causes tidal friction in Earth and vice versa.

tonalite
An igneous rock composed primarily of plagioclase feldspar and quartz with hornblende and/or biotite.

trace element
A minor constituent whose presence or absence does not change the type of rock.

troctolite
A cumulate igneous rock consisting mostly of the minerals pyroxene and olivine.

troilite
A mineral composed of iron sulfide that occurs in iron meteorites.

turbidite
The deposit of a turbidity current, or underwater debris flow.

ultramafic rock
An igneous or metamorphic rock consisting primarily of dark iron-magnesium minerals. *See also* mafic rock.

Universe
The whole known body of objects, energy, and phenomena.

uraninite

A uranium oxide mineral, also known as pitchblende.

vesicles

Gas bubbles in solidified lava flows.

vacuum energy

A repulsive force in the Universe that counteracts gravity. Also called dark energy.

volcanic rock

The rock that forms within or by eruption from a volcano.

volcanism

The processes that result in the formation of volcanoes and volcanic rocks.

white dwarf

A small, hot, dense former star that has exhausted all of its nuclear fuel, is running on leftover heat, and will eventually cool.

X-ray

Electromagnetic radiation that arises from within an atom but outside the nucleus.

zircon

A zirconium oxide mineral.

Index